HUMAN SPACEFLIGHT

Human Spaceflight

From Mars to the Stars

LOUIS FRIEDMAN

THE UNIVERSITY OF
ARIZONA PRESS

TUCSON

To my grandchildren, David, Marisa, and Tristan:
Explore Other Worlds, Protect Ours

The University of Arizona Press
www.uapress.arizona.edu

Printed in the United States of America
20   19   18   17   16   15   6   5   4   3   2   1

ISBN-13: 978-0-8165-3146-2 (paper)

Cover designed by Carrie House, HOUSEdesign llc
Cover image © Igor Zh/shutterstock. Elements of this image furnished by NASA/JPL-Caltech.

Library of Congress Cataloging-in-Publication Data
Friedman, Louis, author.
   Human spaceflight : from Mars to the stars / Louis Friedman.
      pages cm
   Includes bibliographical references and index.
   ISBN 978-0-8165-3146-2 (pbk. : alk. paper)
   1. Manned space flight. 2. Interstellar travel. 3. Space flight to Mars. 4. Outer space—
Exploration. I. Title.
   TL788.5.F748 2015
   629.45—dc23
                              2015005908

∞ This paper meets the requirements of ANSI/NISO Z39.48–1992 (Permanence of Paper).

Contents

Illustrations

Figures

Tables

Color Plates

HUMAN SPACEFLIGHT

Introduction

The thesis of this book is that humans will become a multi-planet species by making it to Mars, but no farther. That is, they will *never* travel beyond Mars. Some find this to be negative—an absolute statement of limits and thus of giving up on the future. My job here is to prove the opposite: that my view of humans exploring the universe with nanotechnology robotics, bio-molecular engineering, and artificial intelligence is exciting and positive and based on an optimistic view of the evolution of both ourselves and our technologies.

Former NASA administrator Tom Paine,[1] a member of The Planetary Society board of directors, had a clear idea about the goal of human space exploration. It was the settlement of Mars. In fact, he hoped we would not find life on Mars so that we could more rapidly devote ourselves to bringing humans there. He wanted The Planetary Society to make Martian settlement our vision and mission. I was more influenced by Carl Sagan and Bruce Murray[2] who, while very supportive of human Mars exploration, maintained their scientific skepticism about the question of settlement. We would argue that we don't know if settlement is possible or sensible; we might "only" explore with outposts and never really settle there (as in Antarctica or in the oceans of Earth). I saw human spaceflight as an ever-increasing faster and farther activity rather than something so destination oriented. NASA had taken that same position with a mantra of "Moon, Mars and Beyond" describing their human space ambitions. President George W. Bush even established the President's Commission on Moon, Mars and Beyond in 2004, which he made into his "Vision for Space Exploration." Ironically (one might say bureaucratically), this became a lunar base program—not only losing the "beyond" but also Mars.

But I have come to realize that Tom Paine was right—settlement of Mars is the rationale for human spaceflight. This is discussed fully in chapter 8. I further have come to realize that Mars not only should be the next goal for humans in space but also is the ultimate and hence only goal, at least physically. Exploring beyond Mars will be done virtually, by processing information from other worlds while our bodies stay at home (albeit, I hope, on a multi-planet home of Earth and Mars).

This conclusion can be phrased as follows: human space exploration will continue *forever*; human spaceflight will stop at Mars. Although this book is about the future of human spaceflight, more than half of it deals with robotic flight—very advanced robotic flight that creates human space travel and human space exploration without human spaceflight. That is not a contradiction—it is just a new way of thinking, a problem perhaps for an older generation but not for future ones where already ideas about connectivity, networking, exploration, and virtual reality influence the perception of "being there." Once we understand that human space travel beyond Mars will be technologically, psychologically, and culturally very different from how it is carried out today and in the near future, then our construct and development of space programs should be different. We can solve the conundrum imposed by the merging of grand visions and political constraints—the biggest problem of space policy.

The reader might now wonder, what about Mars? Why do we say human space travel will change only beyond Mars, and why do we insist that humans still must and will travel themselves to Mars and perhaps even settle there? We will explore this question extensively in the last three chapters of the book. The answer is found in becoming a multi-planet species.

"Space is big, really big," wrote Douglas Adams in *The Hitchhiker's Guide to the Galaxy*.[3] I wrote those same words in my earlier book, *Starsailing: Solar Sails and Interstellar Travel*,[4] before having read Adams's novel because I was struck by the same problem that his Arthur Dent character had—the distances between stars is unimaginably enormous. "How big is space?" you might ask. Well, it might be infinite (we only know what we can see so far), but the observable universe has a diameter of about 90 billion light-years.[5] That equals 850 billion trillion kilometers. That number, incidentally, is more than the number of grains of sand on Earth. In that observable universe there are about 100 billion galaxies, and in our galaxy, there are about 100 billion stars. One of those stars is our Sun, and its solar system has yet to be completely explored—or even have its population fully cataloged. If we divide the number of stars (100 billion × 100 billion = 10^{22}) into the volume of observable universe (assuming it is a sphere since it has no preferred direction), we find about 25 stars per trillion cubic light-years. If that sounds pretty empty to you, then you have it right: space is mostly empty!

The point of playing with these unimaginable numbers is to dramatically illustrate that interstellar travel is a subject of science fiction, not ready for prime time, at least for humans. Even most of the serious technical work, some with brilliant engineering and sophisticated applications of physics, relies on schemes that are fiction—or at least not real in any practical sense. Interstellar travel is discussed in chapter 2.

The barrier of bigness preventing human travel to the stars will be overcome by the enablement of smallness—the evolution of information processing and nanotechnology that will permit virtual human exploration of other worlds. The pathway that takes us there is on light, and the propulsion scheme is to use only light beams and no propellant. How we do this is described in chapters 2–4. In chapter 5, I will introduce the concept of interstellar precursors in the context of the (so-named) 100 Year Starship™ and show which we can start working on today for missions to eventually reach other worlds. Then in chapter 6, I will propose a specific mission to serve as a milestone exiting our solar system, one that might allow us to point at desired interstellar destinations. In chapter 7, I will draw a conclusion that the long-range future of humankind is to extend its presence in the universe virtually—with robotic emissaries, bio-engineered payloads, and artificial intelligence. Payloads will be designed for information processing, not life support. But does that conclusion doom us to be couch potatoes—staying at home forever, confined by the limits of our planet? This is neither culturally nor physically acceptable. Culturally, we remain wanderers and explorers. Physically, the survival of our species requires humankind to become a multi-planet species. We cannot put all our eggs in our Earthly basket. It has too many forces that might cause it to fray—forces such as asteroid impact, large-scale conflict and war, pandemics, global climate change, and other types of environmental destruction, such as destruction of resources, to name a few.

In chapter 8, we find that to become a multi-planet species not only must we humans reach Mars but also it is the only place we can reach. Mars is the only world we can reach on which we imagine human life making it habitable. Indeed, it may have been habitable in the past, and life may have originated there as fast as it did on Earth and even have spread here from there, or there from here. Therefore Mars remains a human destination. It is the laboratory in which the evolution of the human species will be tested and, ultimately, determined. The steps to Mars, from today's robotic program to getting humans there in the 2040s, are described in chapter 9. There we deal with space policy and what is necessary to make getting humans to Mars happen. An appendix on the difficulties and realities of constructing a human Mars mission is included—to simultaneously convince the reader that it is not easy but it is possible. As President Kennedy said, "Do [these] things, not because they are easy, but because they are hard, because that goal will serve to organize and measure the best of our energies and skills, because that challenge is one that we are willing to accept, one we are unwilling to postpone."

Getting beyond Mars (with humans) is impossible—not just physically for the foreseeable future but also culturally *forever*. By the time we make it on Mars, our technologies will have evolved so that the human presence in space is being extended robotically and virtually, rather than corporally—that is, by leaving the human at home. Human presence on Mars simply extends the definition of "home" where we leave the human explorers: home might be the inner solar system where humans can live as a multi-planet species resilient to planetary disasters and extending both physical and mental limits into the universe. It might be imagined as the inner solar system, but most likely it will not be inward from Earth at all, and no farther away than Mars.

These divergent paths into space lead to realizing the vision of spreading humankind outward into the universe and onward into the future. It is to that vision that I hope this book contributes.

CHAPTER ONE

The Human Future in Space

*Let us create vessels and sails adjusted to the heavenly ether, and there
will be plenty of people unafraid of the empty wastes.*
—JOHANNES KEPLER, LETTER TO GALILEO GALILEI, APRIL 1610

This book explores humankind's future beyond Earth. The present and near
future are clouded by politics and the cost of human spaceflight. Although
humankind has successfully left Earth, we have not gone very far. We have
never escaped Earth's gravity field, and in fact, in the 40 years since the
end of the Apollo program, we have not even reached farther into space
than the distance from Los Angeles to San Francisco. If we characterize the
space race as a race between humans and robots, the robots are clearly winning—and their lead keeps widening. Robotic flight technology is advancing exponentially fast, while human flight technology is hardly advancing at
all. With regard to interstellar flight we are like Leonardo da Vinci dreaming
and imagining about flying machines centuries prior to any technology that
can achieve them.

Interstellar dreams of human spaceflight generally imagine transporting
corporeal human bodies and keeping them alive for long periods in the
hostile environment of space. The technology concepts for this are mostly
documented in science fiction literature, and they haven't changed much
in the past 50 years—that it, the whole space age. Those concepts include
massive propulsion systems using technology not yet invented, such as
nuclear fusion or matter-antimatter engines; exotic life-support systems,

such as cryogenically freezing humans and then magically reawakening them with all their functions intact; and generational flight times, such as interstellar arks with long-duration, closed-loop, self-sustaining ecological systems, such as traveling artificial biospheres. These themes appeared in science fiction literature decades ago, and they remain there today. They are not just in the science fiction literature but also in the technical literature. Much of the current technical thinking about interstellar flight follows these same ideas. The concepts are often based on real physics, but not on real technology. The actual technology of human spaceflight is still limited to durations of about one year and distances within the gravitational field of Earth. Not only has the technology not changed much during the space age for human space flight but also even the concepts for it are mired in the past.

On the other hand, spacecraft, information, and biological technologies have evolved enormously in the past 50 years, using discoveries and innovations that were barely imagined when interplanetary flight began. It is common to observe how much more computing power we have in our cell phones and games than we had in the earliest planetary probes or the Apollo spacecraft.[1] Electronics and sensors get smaller and smaller, while information processing and communications capabilities get larger and larger. The field of small spacecraft is evolving from micro-spacecraft (10–100 kilograms) to nano-spacecraft (1–10 kilograms) to pico-spacecraft (under 1 kilogram), and its limits are not known. Limits, except for the speed of light, are also not yet known for communication data rates or for information processing speed and capacity. The combined exponential reduction of electronics size and increases in information processing capabilities give us something like a Moore's law (a systematic and repeated doubling of capability) for rapidly increasing robotics technology, whereas the technology predictions for transporting humans via space flight actually seem to give us an inverse Moore's law, describing a rate of decreasing capability and longer timescales.[2]

Extrapolating from current robotic developments, we will describe super-fast, ultralight spacecraft propelled by lightsails that will enable interstellar travel with the human brain, but not the body. The human brain will be integrated with that of the spacecraft's, utilizing advances even more profound than those in the physical and electronic technologies. These advances are in biomolecular technologies and nanotechnologies and in information processing that can extend the human presence into worlds much farther and faster than we now imagine.

The human in modern space systems can be a virtual explorer of a place, interacting with the place's environment as much as a real explorer but without the baggage of physical transportation. We know this about space exploration because it is already happening in our Earthly environment. Much more astronomy is done by astronomers in their offices or even their homes than by trooping up to mountaintops to be present at the observing telescope. Data reaches the observer at the office or at home only milliseconds later than it would if he or she was right next to the telescope. The facilities for processing the data are usually much better at the office or at home, in addition to the huge improvement in efficiency of time usage.

Modern warfare is also being conducted more and more by telerobotics and by leaving the human operators at a home base. The United States is operating drones in Asia and Africa with high precision and high interactivity while the tele-operators are still in the United States. That might be the best way for humans to explore over interstellar distances (although I hope without targeting weapons on their putative inhabitants). Is it less human to explore and engage remotely? We'll explore this question more in chapter 7, but I will tip my hand now and tell you I think not.

Add to this the increase in robotics and information technologies, the technology of biomolecular engineering, and microorganism manipulation and even more possibilities are suggested. We can certainly send our DNA on long trips out of the solar system. A private space venture sold kits (even in NASA gift shops!) some years ago saying they were going to do just that—with a solar sail spacecraft. Unfortunately, they had no capability in their planned spacecraft for communicating with it or even monitoring it over interplanetary distances, and their design had a very limited lifetime. It was just a stunt supposed to allow those who bought the kits to say their DNA was on the way to the stars. They actually never even launched. (There is nothing wrong with such stunts—or even simpler ones like sending your name into space on microchips or on CD data storage, as we did in The Planetary Society for our members. What is wrong is when false or deceptive marketing is used.)

But just sending our DNA as a souvenir for interstellar destinations is not very satisfying. What if we could program it to interact with a newly discovered alien environment and convert what was sensed into a signal that could then be communicated to us? After all, biology is increasingly being understood as digital, an idea first suggested by the famed quantum physicist Erwin Schrödinger in his book *What Is Life?*[3] A decade before the structure of DNA was discovered, he proposed that life consisted of aperiodic crystals that could be digitally described. Can we program DNA or microbial life in some merging of digital biology and electronics to turn on and off a signal in the spacecraft? That would clearly be an extension of our human presence into the universe. A really far-out idea is that we could alter the quantum state of our spacecraft-carried molecules in a way that a paired quantum state on Earth receives the interstellar probe information instantly (quantum communication). This is wild speculation right now, but the technologies enabling ideas like merging digital biology and electronics are becoming more realistic much faster than are the classical technologies thought of for human interstellar flight.

Space is too big, time is too long, and we humans are too impatient to wait for human life-support developments—we will be ready to use our rapidly advancing technology to take us to the stars virtually long before we invent some yet unknown transport that can make human interstellar flight practical. Human spaceflight in the future will leave the human at home!

I used to be disturbed by this—indeed, I fought against it. In many public speeches that I gave as the executive director of The Planetary Society, the world's largest public space organization, I would assert that we are again in a great space race—this time not between competing political ideologies but between humans vs. robots. I, an admitted human chauvinist, was rooting for the humans. Being human I want humans to advance, not just virtually but really. We are, I would argue, better people if our reach extends our grasp[4] and adventure knows no bounds. I used, and still use, this argument in favor of human spaceflight even when there is so little political will or geopolitical rationale for human spaceflight.

However, I am no longer distressed by the robots winning the ultimate space race—so long as we can have great human adventure, advancement, and victories along the way. That can happen by our embracing the robotic and molecular technologies that will enable humans to go, virtually, to the stars. The experience will be even more satisfying, for we will be able to do so much more, so much sooner than would otherwise have been possible from the old-fashioned ideas of transporting humans over huge distances. It will be not only more satisfying but also more exciting, with adventures of the mind triumphing over adventures of the body. Best of all, I won't have to wait on the invention of "unobtainium" to enable practical human spaceflight to distant worlds—I can hop on the figurative bandwagon of evolving nanotechnology.

Are we thus doomed to remain hidebound on Earth? Emphatically not. Ensuring the survival of our species in the case of some global environmental catastrophe is sufficient reason to compel us to advance human spaceflight to another world. Even in the absence of an immediate threat, our environmental awareness is growing to the point where we should think about the possibility of an uninhabitable Earth in the future (even a temporary one), whether from war, disease, climate change (human or natural caused), or asteroid impact. These possibilities, unlikely and not imminent, are nonetheless real. It is interesting that one of the earliest efforts to advance capabilities of human life support for long-duration spaceflight was in a project called Biosphere 2. (Biosphere 1 is the Earth.) The initial funding for it came from a "survivalist"—a wealthy donor motivated by the fear of human extinction. (Later it became a university research project and lost much of its survivalist motivation.) Fear is a political rationale. So, maybe, is hope—to inspire and motivate our own society and generation (and future generations) with high achievements and new opportunities. These do not come from remaining hidebound on Earth. From such considerations, discussed more fully in chapter 8, we conclude that we need at least one world off Earth that we can reach, if only as a backup. Mars is that world; it is the only reachable planet with atmosphere, water, and an environment on which our human life might be sustained. It is important to understand the concept of "might be"—we still have to prove that it could be.

Many space enthusiasts need no proof; they are confident in their assertion that Mars (and some say even the stark, airless Moon) will be colonized. But I keep thinking of those *Popular Science* covers of the 1950s (and later) showing undersea cities with seaweed farms and mining operations and factories, and a huge sealed submersed structure housing hundreds of thousands, if not millions, of people. That hasn't happened—despite the underwater ocean environment being more human friendly and accessible than space or Mars. Even the Antarctic continent is not colonized—just visited. So, to me, it is not a given we will colonize Mars—only that we have to try. (There are counterexamples, of course—previously uninhabitable areas on Earth, like Las Vegas and Tucson, are now well-colonized due to the creation of artificial environments through the use of air conditioning.)

We have to go to Mars (1) because it is there (and nothing else is!) and (2) to ensure the survival of our species. But, beyond Mars, the nearest habitable world is likely to be at another star system, and the distance from Mars to the stars is enormous. There are no other habitable worlds (for humans certainly and maybe not even for simpler life in our solar system). Astrobiologists and planetary scientists have an increasing interest in the moons of the outer planets (especially Europa at Jupiter and Enceladus and

Titan at Saturn), but only for robotic spacecraft. Humans traveling there would represent an enormously large step, one that defies credulity, especially as we consider the advance of robotics and virtual world exploration. If we think of our first two human space travel accomplishments—Earth orbit and the Moon—and then scale them up in the distance to Mars and then to the stars, we get multiples of distance of 1 (Earth orbit), 600 (the Moon), 375,000 (Mars), 1,000,000 (Jupiter), 2,000,000 (Saturn), and 63,000,000,000 (the nearest star). It's okay to roughly figure the difficulty for human space travel to go with distance—reaching the Moon is 600 times more difficult than reaching Earth orbit (the United States did it 40 years ago and neither they nor anyone else has done it since), reaching Mars is 375,000 times more difficult, and reaching the stars is about 63 billion times more difficult. The scale of space exploration is illustrated in figure 1, with a logarithmic scale—that is, equal distances on the graphic are multiples of 10 in distance.

Maybe we'll[5] see some human reach the Moon again in another 15 years or so, and maybe we'll get to Mars within three decades (an optimistic prediction these days), but the stars are centuries away, if that. By then—and in fact even in the next 10–20 years—the robotic/nano-/biomolecular technologies of which I spoke above will reach the point that human flight to the stars will be enabled virtually so that flying there corporeally will be an antiquated idea—like flying on wings to heaven. Personally, I like the idea of us humans learning how to send our minds to worlds beyond our solar system on the same time scale as learning to send our bodies to Mars.

Between Mars and the stars, I am asserting there is nothing—at least for humans. In part this is because of the hostility and uninhabitability of everywhere beyond Mars, and in part this is because of the time scale mentioned previously. This is even true for the outer planets—too far and too hostile for humans, but comparatively accessible for robots and virtual exploration. By the time we actually attain and utilize our physical presence on Mars, virtual exploration technologies will have taken us everywhere in the solar system and well beyond. Not everyone agrees with this, of course—there are some excellent science fiction books written about humankind settling the solar system and living everywhere on asteroids, moons of the outer planets, Venus, and maybe even Mercury. Some have big ideas about exploiting planetary resources throughout the solar system to build civilizations and colonies on other worlds. I am more constrained—I can't see why or how we would do this nor can I see human evolution of our mind and our accomplishments waiting around for the ability to do so.

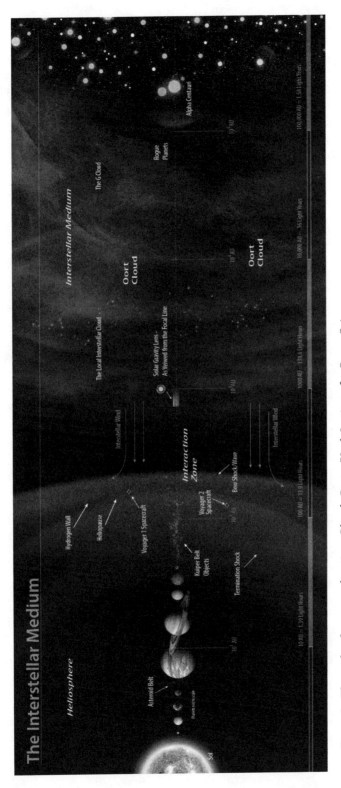

Figure 1. The scale of space exploration. Chuck Carter/Keck Institute for Space Sciences.

Thus, human spaceflight is on two paths—a short one in this solar system, enabling us to become a multi-planet species inhabiting Mars (hopefully as well as continuing on Earth), and the other leaving the humans at home (on Earth *and* on Mars), engaged with our virtual selves on worlds, first in our outer solar system and then about other stars. These paths are realistic, we have already started on them, and the political and technical roadblocks that might inhibit us can be overcome. How we do this is what we now explore.

CHAPTER TWO

Lighting the Path

Solar Sails and Interstellar Travel

Sail forth—steer for the deep water only,
Reckless O soul, exploring, I with thee, and thou with me,
For we are bound where mariner has not yet dared to go,
And we will risk the ship, ourselves, and all.
—WALT WHITMAN

Interstellar flight has begun. The *Voyager* and *Pioneer* spacecraft, two of each, launched in the 1970s, are headed on interstellar flights, however, not practical ones. As of spring 2015, *Voyager 1* was approximately 130 AU from the Sun traveling outward at a speed of about 3.8 AU/year.[1] *Voyager 2* was about 108 AU from the Sun with a speed of about 3.3 AU/year. *Pioneer 10* and *Pioneer 11* were approximately 110 AU from the Sun at 2.5 AU/year and 89 AU at 2.4 AU/year, respectively. While there is no exact definition of interstellar space, it is generally taken to be the irregularly shaped and still uncertain region known as the heliopause where the Sun's influence (in terms of its magnetic field and particles) yields to that of the interstellar environment (see figure 1 in chapter 1). *Voyager 1* has now passed through the heliopause and entered interstellar space. It appears that the *Voyagers*, launched in 1977 on a four-year mission, will stay powered up and in communication with Earth until at least the year 2020, and therefore, at least by one definition, *Voyager 1* is an interstellar spacecraft (flying in the interstellar medium).[2]

The *Voyager* spacecraft have a remarkable story. In the late 1960s, astrodynamics engineers working on trajectories for outer planets missions discovered the possibility of a "Grand Tour"—a once-in-176-years alignment of the large outer planets (Jupiter, Saturn, Uranus, and Neptune) to which a trajectory (orbital path) from Earth could be made to fly by each of them ballistically, that is, only under the influence of gravity without large propulsive maneuvers. They would use the technique of "gravity-assist," whereby spacecraft could fly by one planet and get a boost in velocity, targeting it to another planet or even out of the solar system. The boost in velocity comes about from a miniscule exchange of momentum with the planet about which the trajectory passes. Gravity-assist was used on the *Mariner 10* mission to reach Mercury with the aid of a Venus gravity-assist and on the *Pioneers* to reach Saturn after Jupiter, but no spacecraft had flown with multiple gravity-assists, or to the outermost planets, or on a decade-long trip. So engineers at the Jet Propulsion Laboratory (JPL) began working on a very sophisticated, long-life, autonomous spacecraft with nuclear power to make the journey. This "Grand Tour" project was conceived of to take advantage of this once-in-176-years alignment of the planets. JPL developed the mission idea in the late 1960s, and then it was cancelled (falling under the budget ax) in 1971. I remember it clearly because we were notified about it just days before I was scheduled to present a paper about the mission's navigation requirements, and I had to change my speech to refer to all my work in the past tense. I felt like I was giving a speech on the *Titanic* as it was sinking. But JPL, where I was lucky to be working in an Advanced Projects Group charged with inventing and developing new mission ideas, didn't give up. Our group, led by Dr. Roger Bourke, was known for its innovative ability. Working hard over a weekend, they came up with a back-up plan to use *Mariner*-derived spacecraft (instead of the autonomous, more expensive craft being studied) on trajectories that would just go to Jupiter and Saturn, instead of doing the whole Grand Tour. The spacecraft would be designed for 4-year lifetimes instead of 12-year lifetimes. The trajectories would be designed so that if the spacecraft were lucky enough to live longer, one more planet—Uranus or Neptune—could be encountered. JPL obtained approval for this lower-cost mission—the result was *Voyager* and the rest is history. Thirty-five years later the spacecraft are still working, after having accomplished the grand tour (now without capital letters) and more. Now they are officially on the Voyager interstellar mission. The spacecraft are still making measurements (as I write this), with results showing that the physics of the heliopause is far more complex than we imagined. Even nearly empty space can be interesting.

Although both *Pioneers* and both *Voyagers* are carrying interstellar messages for a hypothetical alien encounter, the earliest star encounter by any of these four spacecraft is more than 40,000 years from now—and those Voyager "encounters" are still no closer than 1.5 light-years from the passing star (a light-year is approximately 63,000 AU or 9.5 trillion kilometers).[3]

Practical interstellar flight is a long way off. By "practical," I mean something that goes to another star system in a generation or two (say, less than 100 years) and something that doesn't merely broadcast our existence to the universe but that also gets back information to its creators. We haven't that capability, or even the knowledge of how to get that capability, yet.

In fact, we currently have only one technology that might someday take us to the stars, and even it needs something we still haven't developed. The technology is lightsailing, and what it needs is a powerful laser beam with high power and large optics to focus light in deep space beyond the solar system.

Only one technology? What about unlimited nuclear energy, matter-antimatter annihilation, warp drives, and diving into wormholes that we read about or see in the movies? Those, unfortunately, still remain as either science fiction or theoretical musings without any technological development, at least now and for the foreseeable future.

Advocates of limitless nuclear energy continue to suggest that eventually engineering solutions will be found to harness the energy. But, all the nuclear energy ideas for interstellar propulsion still require carrying both propellant (such as hydrogen) and fuel (such as plutonium) along in the spacecraft: fission as in atomic bombs, fusion as in hydrogen bombs, or combining matter and antimatter as in nothing that has been invented yet. And when the calculation is made, the amount of propellant required is just too heavy, even if it is converted to energy with higher efficiencies than have ever been invented. While many concepts and even some theoretical designs have been offered in the past 50 years, what bothers me the most about them is how little those concepts and designs have changed over that period of time. Little progress, even conceptual, has been made on nuclear rockets since the first serious effort with the NERVA program (Nuclear Engine for Rocket Vehicle Application) was made in the 1960s. NERVA was a powerful rocket for solar system missions, but "merely" heating gas with fission power would not provide enough speed for interstellar flight. The most serious interstellar designs were from Project Orion, which used continuous hydrogen bomb explosions to theoretically make a pulsed rocket engine—hundreds of thousands of bombs in a single vehicle for interstellar flight! The mass requirements for carrying enough propellant and making enough explosions are impossibly impractical for interstellar flight, although such a rocket would be a powerful vehicle for solar system exploration and applications. The fascinating history of Project Orion is described in *Project Orion* by George Dyson, the son of Freeman Dyson, who was the key originator and advocate of the concept.[4]

The next step up from fusion energy might be getting "pure energy" by combining matter and antimatter in a rocket. Of course we don't know how to do that yet, and even theoretically thinking about how we could contain and control the antimatter to mix with the matter and effectively utilize the resultant energy is beyond our imagination, let alone the state of our knowledge.

A very complete compendium of propulsion methods has been published by the American Institute of Aeronautics and Astronautics and edited by Marc Millis and Eric Davis, propulsion experts in their own right, with contributions from others working on a variety of ideas.[5] Conclusions from this and other surveys of interstellar flight always lead away from rocket propulsion because whatever propellant you imagine using has one limitation: it has mass—a lot of it—that has to be carried over interstellar distances. As noted earlier, the ideas of fusion and antimatter rockets have been around since the beginning of the space age without even a modicum of engineering progress making them more realizable now than 50 years ago. Lightsailing, carrying no propellant and getting its energy externally, seems like the only way we know to propel an interstellar spacecraft.

However, there are at least two other ideas for interstellar propulsion that do not require propellant. One, conceived at the very beginning of the space age, is the Bussard ramjet, which gathers propellant from interstellar hydrogen atoms as it flies. It was proposed by Robert Bussard, a nuclear physicist, who envisioned a large scoop on a starship to gather hydrogen-creating proton-proton propulsion to make a ramjet. But it is only theoretical and is unlikely to really be capable of being engineered. The scoop itself will be a source of interstellar drag. The idea was first proposed in 1960, but no one has made it practical since. A second idea is even more theoretical—creating warp drive (made famous in *Star Trek* as the fictional means of interstellar propulsion). This would be done by creating an energy wave with the spacecraft itself and then using the wave to compress space-time in front of and expand it behind the spacecraft. Creating the wave seems now to be only a theoretical idea—one that requires a negative mass enabled only by a distortion of space-time and an enormous amount of power. But it received some recent attention when Dr. Harold "Sonny" White, a NASA scientist, came up with a design for a small laboratory experiment to test it. Trying such an experiment could enhance our understanding of physics, but it is a long way from creating a technology that can be used for spaceflight. That remains, for the present and the foreseeable future, possible only in science fiction.

Professor Kip Thorne at the California Institute of Technology has a suggestion for interstellar travel that also does not involve propellant: use of a wormhole. To say this is only theoretical is an understatement. Thorne is one of the world's leading astrophysicists and experts in general relativity. A wormhole is a mathematical solution derived in Einstein's relativistic theory of gravity—it has not yet been observed physically. Even though the existence of wormholes isn't proved, their use for space and time travel is a subject of speculation among theoretical physicists. Thorne finds equations that allow faster-than-light travel if the wormhole has specific size and material properties, but those same equations may also allow paradoxical time travel. I personally conclude that it is not possible to stay in the same universe while traveling through wormholes, even while admitting I don't know what is really meant by different universes.

Thus, as we said, sailing on light is the only known technology that can take us to the stars! Solar sailing is the way we do it when near the Sun in the inner solar system, and sailing on focused laser beams (or other electromagnetic radiation) is the way to do it beyond the Sun's reach.

A solar sail is propelled by the force from light's energy. The particles of light, photons, bounce off the highly reflective (made with a coat of aluminum or silver) mirror-like sail. The photons bouncing off the sail transfer momentum to the sail and its attached spacecraft, with a resulting force perpendicular to the flat sail sheet. That light has energy is pretty obvious (just feel it warm up your hand), but to demonstrate its pressure is harder. On Earth we have too many disturbances from the atmosphere and gravity to actually be able to feel the pressure (except in a vacuum laboratory experiment). In tourist shops they often have a spinning windmill in an airtight glass bottle (since there is no wind it actually is a "lightmill"), with the blades of the lightmill white on one side and black on the other. It moves when light is shined on it—supposedly to demonstrate light pressure. Unfortunately, that motion is caused by the differential heating of air particles, heated more on the reflective side (white) than on the absorbing side (black). This is called a Crookes radiometer, and it demonstrates only a thermal effect. Demonstrating light pressure requires a vacuum.

An ordinary sailboat on water operates at the interface of two media: wind and water. The direction of the sailboat is controlled by a rudder in the water, not by the direction of the wind. By steering (tacking) the boat with the rudder, the boat's direction can run with the wind or go almost upwind or anywhere in between. You have to be moving to change the boat's direction. A similar method allows us to steer or tack a sail spacecraft almost toward the Sun. In space the two media are light pressure and the orbital motion derived from gravity. If we tilt the sail so that the force from light pressure adds to orbital motion, then the spacecraft flies away from the Sun. If we tilt it so that the force subtracts from orbital motion, then the spacecraft flies in toward the Sun. In either case the spacecraft follows a spiraling motion—outward or inward—which we can control and hence steer to go wherever we want (see figure 2).

Sunlight is pure energy, and it hits the Earth with a power of 1,370 watts per square meter. Divide the power by the speed of light and multiply by two to account for the incident and the reflected light, and you get the resultant force imparted to the craft. It equals 9.2 micro-newtons per square meter of sail (= 0.18 millionths of a pound per square foot for those of you preferring American units). The speed increase (acceleration) to the sail spacecraft is the force divided by mass ($F = ma$), and thus we have the fundamental design principle of solar sail spacecraft: make the area as big as possible (to collect lots of photons and increase the force) and the mass as small as possible (to convert all that force into acceleration or equivalently all that momentum into velocity). The resultant acceleration is just under 10 micrometers per second per second (9.1 $\mu m/s^2$) multiplied by the area of the sail in square meters and divided by mass in kilograms. That is a little less than one-millionth the acceleration of the Earth's gravity.

The thrust may be low, but it is continuous. Imagine a modest sail of say 100 square m with a 9 kg spacecraft. The acceleration is 0.0001 m/s², which acting over a day increases the speed about 8.6 m/second (there are 86,400 seconds in a day). In 10 days that provides a velocity impulse big enough to move 7,000 km, more than the radius of the Earth. The characteristic of low-thrust spacecraft (be they sails or with electric propulsion creating an ion drive) is that they build up velocity changes slowly but continuously. Chemical rockets provide high-thrust and fast-velocity changes, but brief ones. Low thrust is preferred for long flights.

The numbers cited above are from the power of the Sun at 1 AU (astronomical unit, the distance of the Earth from the Sun). Closer to the Sun it is higher, and away from the Sun it is lower. Once we get to distances beyond Mars, the sunlight power is too low to be useful for propulsion. We have to build up our speed before we get that far out—and we do so by flying in close to the Sun. Sun power is 4 times as great at 0.5 AU (in between Venus and Mercury) and 16 times as great at 0.25 AU. We can get

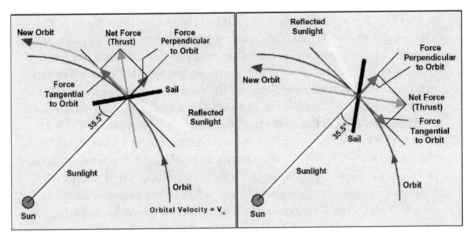

Figure 2. Steering a solar sail. Ares Institute, Inc.

a lot more speed built up if our spacecraft components and materials can survive that close to the Sun. With care, they can.

In addition to "merely" increasing our force from the increased solar power when we fly closer to the Sun, we also are applying the velocity change at the most efficient place in the orbit—namely at the perihelion (closest point to the Sun). Think of an elliptical orbit. That planets and spacecraft move in an elliptical orbit is one of the great discoveries of celestial mechanics (discovered by Johannes Kepler) and great consequences of the law of gravity (discovered by Isaac Newton). The point on the ellipse closest to its focus is called the perihelion when the focus is the Sun, and the farthest point from the focus is called the aphelion. To raise the aphelion (move it farther away), the velocity change is applied most efficiently at the perihelion. (The opposite is true also—to move the perihelion is most efficiently done at the aphelion.) A big velocity change at the perihelion can move the aphelion way out—even to infinity! That is, it can open the ellipse and cause the spacecraft to go on a hyperbolic trajectory escaping the solar system. The big velocity change is what we want from our solar sail near the Sun—it enables us to start out on the interstellar trajectory, or at least on the trajectory escaping the solar system.

Getting a close as possible flyby of the Sun defines the third principle (large area and small mass being the other two) of interstellar spacecraft design with solar sails. That might be counterintuitive—having to go inward toward the Sun to go outward fast—but such are the laws of celestial mechanics.

In chapter 5 we will utilize trajectories that follow these three principles—big area, small mass, and close perihelion—to travel through and beyond the solar system and to define missions that are both interesting in their own right and serve as milestones on the way to the stars. We can achieve velocities high enough to exit the solar system fast, perhaps 15–20 AU/year, and carry out missions to 1,000 AU in perhaps 50–60 years.

As fast and as far as this is, it is not interstellar flight. It is not even as fast as we want for interstellar precursors. As we noted in chapter 1, space is big and the *nearest* star is 271,000 AU from Earth. To reach that distance in 100 years would require speed 100 times faster than even these. We'll discuss what kinds of nano-spacecraft are possible and present some trajectory analysis in chapter 5 that show it would require a spacecraft weighing less than 1 gram per square meter (e.g., a 1 kg spacecraft with a more than 1,000 m^2 sail) going closer than even 0.1 AU from the Sun (less than half the distance of Mercury's orbit) to reach that kind of speed and get to another star system in around 100 years.

Some of my colleagues believe such super-solar sails and super-light spacecraft will happen and that they will carry out meaningful interstellar flight. A more "conventional" idea to keep accelerating the solar sail even beyond the reach of sunlight is to build powerful lasers in space and then focus their beams over huge interstellar distances. Such powerful lasers can in principle be built and even are achievable with today's technical capabilities. But constructing them would be a huge and expensive undertaking, and it is not likely to be done for many decades until space-based applications have several generations of improvement. Robert Forward, a physicist at the Hughes Aircraft Corporation, came up with the idea for laser lightsailing in 1983–1984, publishing both a technical paper in the *Journal of Spacecraft and Rockets* and a novel, *The Flight of the Dragonfly*.[6] Bob was enormously creative and a brilliant physicist who was able to take real ideals from physics and make them into great science fiction stories and to take ideas from science fiction and make them into real physics. His concept of laser sailing is the nearest humankind has yet come to finding a practical means for interstellar flight.

A laser light–propelled sail works like a solar sail, only with a different and highly focused light source. Thus, we still say solar sailing is the technology that can take us to the stars even it will take the space-based, interstellar power laser to implement that technology. (An additional reason that the term solar sailing can apply to the powerful space-based laser-beaming spacecraft is that it is likely that the laser will be powered by sunlight.) Forward's sail in *The Flight of the Dragonfly* (based on his technical paper) was pushed by a

1,500-terrawatt laser, that is, 1.5 million gigawatts, which is approximately 66,000 times the power output of Three Gorges Dam in China, the biggest power plant on Earth. This would have to be put in a solar system orbit to focus the laser continuously into deep space. Forward's laser was powered by a 1,000 × 1,000 km solar array focused through a 100 km lens to a sail 1,000 km in diameter (almost 800,000 square meters). This would enable a flight to the nearest star in just a little less than one century.

In the three decades since Forward's initial work, laser development has come a long way, making the size of a space-based laser somewhat more practical. Professor Philip Lubin at UC Santa Barbara has been conducting research based on these improvements, suggesting designs for space-based lasers applicable to interstellar flight, interstellar communications, and planetary defense (deflecting and vaporizing near-Earth asteroids). He calculates that power systems and laser lens, using phased-array technology, may "only" need to be tens of kilometers in size and 100 gigawatts in power. That is closer to practical, but still a long way off from implementation.

Lasers are not the only form of beamed power (we can think of other electromagnetic radiation besides light); microwave beaming has been strongly suggested by James Benford and others. Very high-power beams using high-frequency radio waves can be achieved—but they cannot be focused like light, and hence their power is lost as the beam disperses into space. Recently, Benford organized Project Forward,[7] a project of the Icarus Interstellar organization, to advance beamed propulsion for interstellar flight.[8] Forward's idea lives on in another new project, Project Dragonfly, a project of the Initiative for Interstellar Studies (i4is).[9] Magnetic sails, where a large superconducting coil generates a spacecraft-generated magnetic field that deflects the solar wind (not sunlight but the photons and electrons from the Sun), have also been described. The deflected solar wind acts like reflected sunlight. The force from the particles is much less, but theoretically the coil generating the magnetic field is much lighter than the sail used to reflect light. A practical way to do this has not yet been devised. For now the only technology we have is to fly on sunlight (not even laser light yet).

We have reviewed all the ideas for interstellar flight in this chapter and find ourselves waiting (wishing) on inventions such as fusion engines, antimatter machines, warp drives, wormhole travel, or the enormous engineering of a giant laser-in-space. Contrary to all these, solar sailing exists now, and it can enable flight outward from the solar system indefinitely. How fast and how practical depends on how small we can make the spacecraft, how big and light we can make the sail, and how close to the Sun we can fly. In the next two chapters we will explore today's solar sails. Then we will examine the engineering questions of nanosats (chapters 3 and 4), and in chapter 5, we will propose a series of steps that we can begin to take today toward interstellar flight. One potentially interesting step and milestone might be a mission to the solar gravitational lens (chapter 6). We will find that the future of all travel beyond the solar system is robotic and virtual — no humans will be carried on such journeys (chapter 7). Realizing that, we can then turn to what the future of humans in space is and earnestly seek our destiny as a multi-planet species on Mars (chapters 8 and 9).

CHAPTER THREE

Sailcraft in Turbulent Political and Technical Waters

If you want to build a ship, don't drum up people to collect wood and don't assign them tasks and work, but rather teach them to long for the endless immensity of the sea.
—ANTOINE DE SAINT-EXUPÉRY

I was first introduced to solar sails in the mid-1970s when I was at the Jet Propulsion Laboratory (JPL). An orbital analyst named Jerome Wright, then working at the Battelle Memorial Institute, discovered a solar sailing trajectory that rendezvoused (matched both position and velocity) with Comet Halley. This was remarkable—because Comet Halley flies in a large eccentric orbit through our solar system, not just far from Earth but opposite relative to the orbits of all the planets. Since it is going retrograde, its angular momentum is hugely different from Earth's, and it takes an enormous amount of energy to match its position and velocity.[1] The aerospace and scientific communities thought it impossible to find a rocket or any propulsion big enough to be practical for a rendezvous during its twentieth-century (1985–1986) apparition.[2] But Wright was clever—he (actually his trajectory) spiraled in toward the Sun and then used the larger solar power and the efficient orbit perihelion location to flip over the orbit, raising its inclination higher and higher until it was greater than 90° and finally reached 160° to match Halley's. The Comet Halley rendezvous required a huge sail—640,000 square meters (a half mile on a side), necessary to carry the heavy spacecraft weighing nearly 1,000 kilograms—the state of the art

Figure 3. Comet Halley spacecraft: heliogyro. NASA/JPL.

in the 1970s. The whole idea was beyond state-of-the-art, too audacious not only for its time but also for ours. To illustrate just how audacious our thinking was, we decided to replace the square sail shape with a spinning heliogyro sail design consisting of helicopter-like, thin, narrow blades 7.5 kilometers long! That is, a spinning structure 15 kilometers in diameter (if it was hovering over New York City, it would cover Brooklyn). We did that because the engineers and mathematicians found the dynamics of the spinning object (even one that long) easier to analyze than the so-called stable square sail—stable but with very uncertain dynamical behavior.

In the mid-1970s, we were capable of bold and audacious ideas. The Grand Tour was about to be undertaken,[3] a journey that would allow the accomplishment of exploration missions to all the planets of the solar system in the twentieth century.[4] We had just completed the first successful landings and mission operations on Mars and Venus, and we had sent 24 men to the Moon and conducted two automated sample returns and two automated rover missions there as well.[5] The "can-do" philosophy of the space age was strong.

Figure 4. Comet Halley spacecraft: square sail, flying interplanetary trajectory, view of back side. NASA/JPL.

Within a few years, that philosophy was replaced by "can't do." Not only was the solar sail idea too big but also anything associated with it ended up on the floor. NASA "banned" the solar sail from discussion, even from technology studies. Its competition to power the Halley rendezvous mission, ion drive technology (solar electric propulsion), had strong bureaucratic and program support in NASA and won over solar sailing in a rather exciting "shoot-out" for the idea of a Comet Halley rendezvous mission. It was heady stuff, but no amount of low thrust (whether from ion drive or solar pressure) could overcome political forces driving down the space program as the Project Apollo era ended. Ion drive won, but Comet Halley lost—the United States ended up with no mission of any kind, not even a simple fast flyby. By way of contrast, the Soviet Union, Europe, and Japan all conducted audacious first-of-their-kind missions to fly through Comet Halley.

Voyager and *Viking* were Apollo-era decisions. Already post-Apollo, the Nixon administration had rejected even finishing the Apollo program (which had scheduled two more lunar landings)[6] as well as a national commission's (headed by Vice-President Spiro Agnew) recommendation for a human Mars follow-up, deciding instead that NASA could continue to exist but that they could have no mission. They could just develop a new vehicle (the space shuttle), but not designate any destination for it. The Nixon administration's lack of support for space (even the initial shuttle-only policy had its budget cut in the immediate years after it was approved) was followed by the Carter administration's declaration of "no high-challenge [space] engineering initiative." Ironically, the Carter administration failed to recognize that the shuttle was already a high-challenge engineering initiative and decided to put all the U.S. space eggs in its basket—all civil launches were to go on the shuttle. Not to be outdone in negatives, the Reagan administration floated a plan to eliminate planetary exploration entirely from NASA's program (including a proposed mission to Comet Halley).[7]

But even while political support for planetary exploration and budgets for space technology were plummeting, we (those of us in the trenches trying to make missions happen) were still sure that the United States would not ignore Comet Halley's 1986 apparition and would at least send a simple flyby mission there. The rendezvous idea (matching the comet's velocity as well as intercepting it) was beyond our means, but intercepting it for a very fast flyby was well within our capability—and actually rather easy. In fact, Europe and Japan, neither of which had ever attempted any planetary mission, each announced they were doing a Comet Halley mission. And the Soviet Union broke new ground by re-planning their 1984–1985 Venus mission to swing by Venus and use a gravity-assist there to fly directly into the comet! The Reagan administration's approach (sold to them by some naïve and a little bit kooky staff ideologues) was to get the government out of the way so that private companies could do it—and make money from the video rights.[8] They did get the U.S. government out of the way, but fortunately Europe, Japan, and the Soviet Union continued their missions.

Seeing this coming, in 1978, I came up with an idea to organize the International Halley Watch (IHW), hoping it would generate enough momentum to overcome the political shortsightedness and allow us to create an American mission in collaboration with the rest of the world. The IHW was successful in harnessing and utilizing resources around the world, although it did not meet my personal goal to make a U.S. mission happen. We did, however, create opportunities for and enable American contributions to the global exploration of this once-in-a-lifetime event.[9]

I didn't realize it at the time, but the IHW global effort was a bit like the global effort Halley himself initiated, not with "his" comet but with the Transit of Venus in 1761/1769. Halley died before the transit, but in 1716 he recognized the important scientific opportunity that this particular rare celestial event would provide (it happens slightly less frequently than once per century). He urged global cooperation among astronomers, politicians, and explorers to enable simultaneous observations to be conducted at disparate locations around the globe—and thus provide measurements that would enable calculating the then-unknown size of the solar system. Essentially he proposed a Halley watch—of Venus. Twenty-eight years after his death, it was successful.

The International Halley Watch was also successful. It brought ground-based observers all over the world together with space scientists in the Soviet Union, Europe, Japan, and even the United States (and other countries) to provide coordinated and the first close-up views of the comet. As a mission analyst at JPL in 1979–1980, I helped identify the mission opportunity (together with others at JPL, in the Soviet Union, and in France—notably Professor Jacques Blamont) that allowed the Soviet Union to re-target their intended 1985 mission for Venus to be a Venus-Halley[10] mission and fly right through the comet close to its nucleus. The United States contributed navigation information to the Soviets and both contributed navigation information to the Europeans, enabling highly accurate flybys to be performed by the Soviet and European teams. This was done during the time of Cold War hostility and indeed in an atmosphere of official government hostility. Official meetings weren't permitted—we all participated unofficially. Scientific cooperation triumphed, much as it did in the eighteenth century when astronomers from around the world were able to coordinate observations of the transit of Venus even while Britain, France, Prussia, Austria, and Russia were fighting wars in Europe (and a bit in the then-nascent America). I like to think that our scientific cooperation in the mid-1980s at Comet Halley contributed a (perhaps only a wee) bit to the emerging global cooperation that occurred a few years later with the end of the Cold War.[11]

The Comet Halley exploration mission story is a bit of an aside in this book about the future of human exploration, but it emphasizes a central point of the book—viz. that human exploration and human engagement need not involve traveling to the targets. In 1761 and 1769, we had to send our humans on far-flung voyages to distant lands to make their measurements. It was definitely human exploration. In 1986, we kept the humans at home and sent our robots out to make their measurements—with exactly the same kind of scientific cooperation and organization and with the same feeling of human accomplishment and significance. To dramatize this analogy further, let me quote Andrea Wulf's telling of the story in *Chasing Venus* of international cooperation in 1761 for the transit measurement:[12]

> Across the world scientists busied themselves with last minute prepara-
> tions, united in their endeavor no matter what their nationality or reli-
> gion, no matter if they had travelled thousands of miles or had stayed at
> home, no matter if they had a 25 foot telescope or a hand-held tube, they
> were all striving towards a common goal. In the midst of the seven years
> war the astronomers had overcome national boundaries and conflict in
> the name of science and knowledge. Now with only a few hours left
> until the transit they could do nothing more than hope that the weather
> would be on their side.

Things were just the same 225 years later when scientists across the world
sat by their science instruments and data terminals overcoming Cold War
conflict in the name of science, waiting and hoping that their spacecraft
would work just a few more hours—into the comet.

The Comet Halley endeavor, international cooperation with robotic
spacecraft, was also human exploration. It was not only, of course, an
intensely human story with personalities, language, art, and even religious
connections but also a story that had humans at the controls and humans
receiving the data. The exploration was even more human than what space
agency bureaucracies label "human exploration," when all they mean is
human pilots or human passengers. Understanding that humans explore
even when they are not physically present on the place they are exploring
will help us prepare for the future of human exploration.

As noted, solar sailing was "banned" from NASA after the demise of
our Comet Halley rendezvous idea. Two private efforts—presaging the
time of private spaceflight by two decades—were organized, the World
Space Foundation (WSF) in the United States and the Union pour la
Promotion de la Propulsion Photonique (U3P) in France. WSF succeeded
in manufacturing its own solar sail and came very close to getting launch
space on the space shuttle for a deployable sail. However, opportunities
for shuttle and secondary payload launches disappeared in the wake of the
Challenger shuttle accident. In the late 1980s, a promoter by the name of
Klaus Heiss came up with the idea of a solar sail race to the Moon during
the five-hundredth-anniversary year of Columbus's voyage to America. The
Columbus 500 proposal (and promises of cash prizes) enticed a number of
student and amateur groups, and a few semiprofessional ones, from around
the world into paying an entry fee and developing individual solar sail

designs to enter the competition.[13] The ideas were good, but the competition had few sponsors and never materialized.

Just as ocean-sailing ships on Earth have many different shapes and designs, so too do solar sailcraft. Earlier I cited the two basic principles of solar sail design: large area and low mass. For spacecraft built on Earth (which is all we can do for now), this means a deployable structure. It was the advent of the shuttle with its large cargo volume that first made this possible for designs in the 1970s, and we will review some of those designs in a moment. Now, with the advent of smallsats and nanosats, deployable structures can be packed in smaller volumes to fit on conventional (expendable) launch vehicles and even as secondary payloads piggybacking on a larger spacecraft launch.

At JPL for the Comet Halley mission, we considered three basic shapes for a sailcraft: square sail (looks like a kite), a spinning sail (looks like a disc), and the heliogyro (looks like helicopter blades). The major criterion then, and for the most part even now, for selecting a design is ease and reliability of deployment. Another key factor in the design is how to control the spacecraft attitude (which way it points), that is, its roll, pitch, and yaw. The square sail is the most conventional, and the most straightforward way for its deployment is with extendable booms (either mechanically unfurling or inflatable). Tip vanes (small sails at the end of the booms) are a way to control the attitude—using the vanes to create differential torque in any of the three directions required. The Comet Halley solar sail spacecraft design was a heliogyro, described in my earlier book. That book and others written afterward about solar sails are cited in the references.[14]

I noted earlier for the Halley Comet rendezvous that we started out with a square sail design using extendable booms and tip vanes but ended up choosing the heliogyro design based on our greater confidence about its dynamical analysis during deployment. Spinning is a great asset to deployment, and the helicopter-like blades were "easy" to twist (with a motor at their base), which allowed differential torque to be created by changing the sail area. The world's first solar sail spacecraft,[15] The Planetary Society's *Cosmos 1*, utilized a three-axis stabilized design (like square sails) with triangular sails enabled by an inflatable boom deployment. The triangular sails permitted it to use motors at their bases to torque them like heliogyro blades for attitude control.

In 1990, the Soviets deployed an unfurlable disc called *Znamya* (*Banner*) from their *Progress* spacecraft used to ferry supplies to their *Mir* space station. Once deployed it was observed from the *Mir* and could also be seen from Earth. Although it looked like a solar sail, it was designed only to reflect sunlight as a space mirror. The Soviet goal was to create large solar reflectors that could be used to light dark areas during the long Siberian winters.

The first solar sail flight was accomplished in 2010 by Japan with their spacecraft, *IKAROS—Interplanetary Kite-Craft Accelerated by Radiation of the Sun*.[16] Their mission has now been officially accepted into the Guinness Book of Records as the first solar sail flight. By hitching a ride on an interplanetary mission, the Akatsuki mission bound for Venus, it was able to deploy and fly without having to make rapid tacking maneuvers as would be required in Earth orbit. It flew in deep space as would a sailboat on the open ocean. *IKAROS* was a brilliant mission plan conducted as a technology experiment piggybacking onto another mission by a very small group in the Japan space agency. The spacecraft weighed more than 300 kg with a 400-square-meter-sail, yielding a rather low acceleration for practical solar sail flight. But because it was already on the interplanetary trajectory, it could fly with only light force for propulsion and attitude control.

IKAROS was very different from our *Cosmos 1* and all other previous solar sail designs. The Japanese chose a spinning approach for deployment of their successful sail (although it was neither a disc nor a heliogyro but more like the petals of a flower). It was spinning for its deployment but then used a very clever 3-axis control of its attitude by changing the reflectivity with imbedded LEDs on different parts of the sail. The differences in reflectivity created differences in pressure and hence a torque that could be controlled by electrical on-off control of the LEDs. The Japanese space agency interest seems not to be in pure solar sailing but in using solar sails in conjunction with electric propulsion for a hybrid design providing power and propulsion for outer planet missions.

NASA's first solar sail was to have been *Sunjammer*, scheduled to fly in 2016. But the project was cancelled in 2014. *Sunjammer* is a square sail, as was NASA's earlier *NanoSail-D* (which did not fly as a solar sailor), with tip vanes for attitude control. The sail, being built by the L'Garde Corporation in California, was to fly in interplanetary space to and perhaps beyond the Sun-Earth L1 point—a point of stability in the combined gravitational fields of the Earth and Sun. The L1 point is located between the Earth and Sun, and it presents a vantage point to monitor solar storm output and provide advance warning to Earth before the particles reach the Earth and cause disruption in our ionosphere. Eventually spacecraft even closer to the Sun will permit longer warning times—which is why the *Sunjammer* had the goal of going beyond L1. This represents one of the main early mission applications for solar sails. Their spacecraft, intended as a NASA technology flight demonstration, was larger and was to be more completely instrumented than the CubeSat-based spacecraft. It would have used 5-micron kapton for the sail material and would also have tested the operation of vanes in the corner of the sail for attitude control. The planned spacecraft mass was 55 kg with a sail area approximately 35 × 35 m. If they achieved this, it would have had an impressive acceleration of 20μg (20 millionths the force of gravity, 6 times that of *Cosmos 1*, 3 times that of *LightSail®*, and 18 times that of *IKAROS*). This is not enough for practical interstellar precursors, but it was a promising step, as we expect spacecraft mass to decrease in the future.

In the next chapter we'll discuss nanosats and CubeSats, including The Planetary Society's *LightSail®*—they are square sails with mass less than 5 kg. Deployment becomes much easier when the spacecraft is small. Since they so far have been designed for use in Earth orbit, and hence within the Earth's magnetic field, their attitude control can be accomplished with torque rods and momentum wheels. However, this cannot be used in the long run, since sails are not useful for making many maneuvers in low Earth orbit—like sailboats on Earth they want to sail in the open ocean and not in harbors.

The two private companies cited above that built solar sails, WSF and U3P, also chose square sail designs that were 3-axis stabilized controlled. In paper studies by ad hoc groups, student projects, etc., a number of other designs were conceived: umbrella-like deployment, a fedora hat spun out of a rocket, and flower-like designs with unfolding petals, to name a few. The complications from having to package the large flat sail into a container for rocket launch, and then having to deploy it reliably in space, may be temporary, owing to the necessity of building space vehicles here on Earth. In the future it is possible that sails will be manufactured in space. After all, they do not require a lot of material, and the zero-gravity, vacuum characteristics of space lend themselves to an easier handling of gossamer structures. If a plastic or carbon-carbon fiber substrate can be manufactured in space and then aluminum or silver particles deposited on it, we would be able to make large and very lightweight sails. Techniques and materials to make the substrate dissolve or disappear (e.g., by the plastic materials subliming) exist, and if they could be used, the resulting sail would be only the angstroms of metallic particles. This might be the sail that could achieve interstellar speeds. No research on this has yet been done—but it might offer the only way to achieve interstellar flight speeds without propellant or a beamed energy source.

Table 1. Solar sail designs

Name	Configuration	Deployment	Attitude control	Flight record
Comet Halley (JPL)		Centrifugal force on spinning spacecraft	Twisting of the blades, creating torques	Never built
Znaimye (Russia)		Centrifugal force of carrier spacecraft (progress)	None— no motors	Built and flown in low orbit (with atmosphere) and without control
Cosmos 1 (The Planetary Society with Russia)		Inflatable booms pulled the sail out	Twisting the triangular sail blades	Built and launched; launch vehicle failed
IKAROS (Japan)		Unfurling and stages of spin and despin deploying the sail; no booms	On-off LEDs varied the reflectivity in different areas, creating torque	Achieved first solar sail flight; flew in interplanetary space

Table 1. Solar sail designs (*continued*)

Name	Configuration	Deployment	Attitude control	Flight record
Sunjammer (NASA with L'Garde Corp.)		Inflatable boom extend sail	Control vanes at the end of booms	Originally scheduled for launch in 2016; now cancelled
LightSail® (The Planetary Society)		Unrolling of mechanical booms	Momentum wheel and torque rods	Spacecraft built; awaiting flight; deployment test flown in 2015

Table 1 shows the very different designs of the solar sail spacecraft to date (2015). The driving factors on the design—the method of deployment and the method of pointing control—are listed. I had the pleasure of working on three of those craft—the JPL heliogyro and The Planetary Society's *Cosmos 1* and *LightSail*®. The heliogyro development was discussed in my earlier book (see fn. 14); in the next chapter we'll look at the two Planetary Society craft.

CHAPTER FOUR

The Planetary Society and Solar Sailing

We have lingered long enough on the shores of the cosmic ocean.
We are ready at last to set sail for the stars.
—CARL SAGAN (*COSMOS*)

Carl Sagan, Bruce Murray, and I formed The Planetary Society in 1980 to prove and harness the popular interest in planetary exploration—a great government enterprise that was in danger of being discontinued. The means to prove public interest was a membership organization where interested people could both get information about and feel involved in planetary exploration. We produced a beautiful magazine, *The Planetary Report*, with the latest pictures from other worlds and well-written popular articles by scientists involved in planetary exploration. In the pre-Internet age, access to firsthand information and images from space missions were hard to get and not generally available. When our business consultants told us that membership dues alone wouldn't create much of an organization and that we had to have worthy projects that could both engage public interest and attract large donations, we responded immediately with the idea that we could fund novel experiments and new ideas that would seed the big projects of the future. Our first project was private funding for the search for extra-terrestrial intelligence (SETI), required after the U.S. Congress belittled NASA's program and cancelled its funding. Another early effort was to fund a technique to discover extrasolar planets—more than a decade before the first one was discovered. We also funded a pilot demonstration of in situ propellant production for a putative Mars sample return mission and tested a novel balloon concept for flying long distances at low altitudes on Mars—all in the early 1980s.[1] All of these early projects were Earth based. I remember one conversation with Sagan in which he mused about whether it would ever be possible to actually create our own space-based project, an experiment onboard a spacecraft, or perhaps even our own mission pushing the technical and discovery limits for space exploration. We thought it might be possible—and by the 1990s it was. In that decade, we managed to get a small radiation-measuring payload on the *Pathfinder* mission landing on Mars and a microphone on the Mars Polar Lander, which sadly crashed on the Martian surface in 1999. These were the first two privately funded experiments to go to another planet.

Sagan and Murray were also solar sail enthusiasts. Carl wrote a hugely popular article about solar sails in *Parade* magazine, the most widely read magazine in the United States. Bruce led advocacy for their development when he made them and the Halley Comet mission possibility one of his purple pigeons[2] at JPL. And, of course, I was an enthusiast—after all, as Bruce would keep saying, "He wrote the book." But, as visionary as we could be, we were mission realists and were not into chasing pipe dreams just for the love of technology. We saw no opportunity real enough to push our own role in solar sailing.

Carl tragically and prematurely died in 1996. In 1999, our Russian colleagues came to us with an offer too good to refuse. They could provide a free launch and a very-low-cost approach for a solar sail development piggybacking on an inflatable structure they were building for a re-entry shield. With Bruce's support, and that of The Planetary Society's board of directors, we assembled our own U.S. team to critically examine and develop their idea and produce a cost estimate that we hoped might be affordable and appealing to the members of the society. Thus was born *Cosmos 1*—the world's first solar sail spacecraft.

The free launch was a result of the post–Cold War desire to use (and perhaps commercially market) intercontinental ballistic missiles (ICBM) (stripped of the nuclear bomb payloads) to launch satellites into orbit. The Russians offered us the *Volna* (*Wave*), a submarine-based ICBM to take a solar sail spacecraft to orbit.[3]

Our budget estimate (approximately $4 million) was large for The Planetary Society (our previous biggest projects, SETI and Mars balloons, were under $1 million), but small by space agency standards.[4] But in addition to this being a fortuitous time with the offer of the free launch and for Russian-American cooperation, it turned out to be a fortuitous time for another reason: it was during the peak of the Internet "bubble." Money was growing on metaphorical trees in the Silicon Valley and elsewhere among Internet startups. Two entrepreneurs with Caltech roots, Bill and Larry Gross, created a start-up incubator called Idealab near our home in Pasadena. One of Idealab's early ventures was a space company called Blastoff![5]

Blastoff! provided initial funding for a feasibility study of the Russian-launched, Planetary Society–managed, Earth-orbiting solar sail project. After an intense review in Russia, we were convinced that the project was feasible and practical. We came back from our Russian review trip all stoked up to go ahead, only to find that Blastoff! had hired a new CEO, Peter Diamandis,[6] who found the solar sail consideration too much of a diversion from their private lunar lander effort. He might have been right—we were selling the solar sail project to them as a low-cost opportunity for a reasonably fast (3–4 years) space success while they tried to build up the much harder and bigger effort for the lunar mission. He had his hands full—in fact, too full. Blastoff! folded in 2001.

Another Internet entrepreneur, Joe Firmage, became very interested in our solar sail while he was creating a science-motivated Internet portal venture. Joe had a strong interest in astrophysics (including a self-described alien encounter experience) and a desire to support scientific inquiry into the questions of space exploration and the search for extraterrestrial life. His new venture provided funding for The Planetary Society to support the creation of SETI@home, a popular early citizen-based venture for the public to help analyze radio astronomy signals, searching the skies for possible extraterrestrial life. Joe then teamed with Ann Druyan, the widow of Carl Sagan, who was developing her own production company, Cosmos Studios. Ann, of course, had a strong interest in the solar sail and strong support for The Planetary Society. She and Joe saw the solar sail project as a great opportunity for their new partnership to create a breakthrough accomplishment and story for a Cosmos Studios media documentary. They committed to the solar sail venture led by The Planetary Society.

In honor of Ann and her new venture, and with a historical tip-of-the-hat to her and Carl Sagan's extraordinary television series, *Cosmos*, we named our mission *Cosmos 1*. The Russians had their own series of *Cosmos* spacecraft going back to the beginning of the space age,[7] and they did not like our name, but we had to point out to them that no one actually owned the name Cosmos, and so we stuck with it. The "1" we stuck on because we intended this to be the first spacecraft to achieve solar sail flight. In particular, that meant we had to fly in a high-enough orbit where atmospheric forces were negligible, that is, above 820 kilometers.[8] What made our first-ever solar sail flight possible was being able to get a launch to this orbit, piggybacking on both the Russian military desire to convert ICBMs into commercial launch vehicles (a figurative "swords into plowshares" post–Cold War dividend) and the Russian aerospace industry development of inflatable structures for re-entry vehicles.

The Planetary Society had a long working relationship with the Space Research Institute (IKI) of the Russian (formerly Soviet) Academy of Sciences and with the Lavochkin Association, a large aerospace company in the Russian space agency responsible for, among many other things, planetary spacecraft. Lavochkin and IKI had carried out all the Soviet planetary missions, and we worked especially closely with them during the development of the Venus-Halley (VEGA) mission, Phobos 1988, and Mars 1996 projects. Lavochkin was now looking to develop commercial applications, among these an inflatable re-entry shield that could be taken to orbit and used to de-orbit old satellites or other orbit debris. Their inflatable design could be used to make inflatable booms that would deploy a solar sail.

Cosmos 1 was a clever design: eight triangular solar sails deployed and held rigid by inflatable booms (approximately 15 meters long). Motors at the apex of the sails on the spacecraft's central hub allowed the sails to be individually tilted (actually in pairs with four motors, each controlling two of the triangular sails), which allowed the sail to be controlled in roll, pitch, and yaw and thus the spacecraft to be turned in any direction so that it could track the Sun.

The spacecraft development was primarily Russian although The Planetary Society provided expertise to oversee the project, contribute to the system design, and assist on particular subsystem requirements, such as the spacecraft radio. Jim Cantrell, Harris M. "Bud" Schurmeier, and I led The Planetary Society team, with the participation of other consultants from time to time. Cantrell had been working with The Planetary Society since his student days at Utah State University and had helped develop the Mars balloon before working on military and civil space programs in the aerospace industry. Schurmeier was one of the great NASA project managers, having led the *Voyager* and *Galileo* missions before becoming associate director at the Jet Propulsion Laboratory.

The Russians promised a two-year project schedule—which we knew was optimistic. It turned out to be five years, with more than one year lost due to what we thought was a free extra bonus test flight opportunity. In the first year of development, the Russians offered to build a scaled-down model of just the sail and its deployment system and fly it for free on a suborbital flight they were using to test their re-entry shield. We rushed off

Figure 5. *Cosmos 1* sail spacecraft. The Planetary Society.

to Best Buy (a consumer electronics store) and bought some video cameras to mount on their spacecraft so we could video the deployment while at the apex of the suborbital flight. The mission was launched on a *Volna* rocket out of a submarine from the Barents Sea, and the flight was to end in Kamchatka, where the capsule and the video would be recovered. Unfortunately there was an anomaly with the rocket (a euphemism for flight failure), and nothing was ever recovered. I, however, gained a unique experience, being able to travel into the Barents Sea on a Russian navy ship (sailing out of Murmansk) to observe the submarine ICBM launch. A Cosmos Studios

Figure 6. Author on ship observing Russian launch of solar sail test flight. Photo by MPH Entertainment for Cosmos Studios and The Planetary Society. The Planetary Society.

video crew accompanied me, gathering material for the Cosmos Studios's planned documentary, and I provided real-time live commentary from the ship to The Planetary Society via satellite phone (see figure 6).

On this same trip I saw a nice demonstration of public interest in the future of human spaceflight. On the day before our attempted launch, we were walking around the city of Murmansk. At a kiosk on one of the main streets, I noticed a newspaper with an article about our upcoming solar sail launch. I bought the newspaper, and turning the page, I was delighted to see what I consider the best picture ever capturing the vision of solar sail flight (see figure 7). We've never been able to track down the artist or where he or she might have gotten his/her inspiration. The way it represents human spaceflight is why I like it so much.

The suborbital flight was attempted in July 2001. It took nearly a year to get a clear understanding of what had gone wrong with the *Volna*. Meanwhile, we returned to the building of our spacecraft—the test flight had been a diversion, and while either a successful or unsuccessful deployment would have been helpful to our engineering, the nil result from the rocket failure turned out to be a time delay and diversion of no help.

Figure 7. Depiction of solar sailor, from *Five Corners,*
16–22 July 2001, page 6, Murmansk, Russia, newspaper.

The spacecraft was finally built and readied for launch by the beginning of 2005. After final tests, reviews, making arrangements for worldwide
ground station tracking and telemetry reception, and organizing mission
operations, we set up a satellite control station in our Planetary Society
office to maintain close contact and shared operations with the in-house,
semi-secret one at Lavochkin (used for both military and civil satellites).
Then, we scheduled the launch for June 21, 2005.

For this launch, I elected to stay in Moscow and go to the mission control center at Lavochkin with Bud Schurmeier. There, together with my longtime friend and colleague, Slava Linkin from IKI who more than anyone organized the Russian effort to make this happen, we watched and witnessed the launch communications provided by the Russian armed forces responsible for the *Volna* launch and the initial radar tracking of the spacecraft to orbit. Jim Cantrell stayed at The Planetary Society to run the society's mission operations activity and to maintain liaison with our members and the press.

On June 21, 2005, *Cosmos 1* went into the sea when the first stage of the *Volna* failed 82 seconds after launch. We had produced and delivered the world's first solar sail spacecraft, but it never got its chance to tell us how it would work.[9] I was pretty disheartened. But The Planetary Society received many plaudits for a great effort—not just from the space community and world press but also from our members who had donated directly to the project. With strong member support we made a commitment to not give up.

Our board of directors and particularly Ann Druyan also remained strong supporters, urging me to not give up and expressing confidence in our abilities to recover. Ann, with her entrepreneurial interest in Cosmos Studios, tried hard with me to find new sponsors and donors. It seemed a natural for media companies, and she was successful in getting seed funding from the Discovery Channel—which enabled us to investigate several ideas ranging from a *Cosmos 1* repeat to new smaller spacecraft with lower costs and risks. But the Discovery Channel funding was only seed money, and no media deal could be made to continue their support. Finally, after four years, another serendipitous confluence of funding and technical opportunity emerged to create a new solar sail project. This became *LightSail*®. The technical opportunity originated with NASA's *NanoSail-D*[10] spacecraft—a low-cost, creative skunk-works project of two NASA Centers: the Ames Research Center and Marshall Space Flight Center. It had suffered the same fate as *Cosmos 1*: its launch failed, on the maiden flight of the new commercially developed *Falcon* rocket, in August 2008 and the spacecraft was lost. The NASA program that created *NanoSail-D* was shutting down, but they had a spare spacecraft, and they offered it to The Planetary Society if we would arrange for its launch.

We said yes, but after six months of talk, NASA could not take yes for an answer. They found another partner with the U.S. Department of Defense, building a satellite called *FASTSAT*. However, by that time Jim Cantrell (the leader and system design manager for The Planetary Society solar sail team) and I were impressed with the ultra-lightweight design of *NanoSail*.

We got together with Tomas Svitek, who ran a small company with a big name, Stellar Exploration, Inc., and who was an expert on nano-spacecraft design. After some consideration, we believed that we didn't actually need the spare *NanoSail* spacecraft but that we could make a new and improved one of our own. Ours, unlike *NanoSail-D*, would have a control system, two-way radio communications, and onboard imaging cameras, and it would be launched to a high-enough orbit with negligible atmosphere to impede on solar sail flight. None of these attributes were in the *NanoSail-D* project. The lightweight, smaller design, which they pioneered, made it lower cost, more affordable, and more practical than *Cosmos 1* to attempt the world's first solar sail flight with private funding. It was an extremely innovative idea and one that led to a new generation of low-cost spacecraft for special-purpose missions. We could piggyback on the emerging CubeSat industry and on secondary launches from various candidate rockets around the world. Jim Cantrell, Tom Svitek, and I created a new solar sail team to make the attempt. This became *LightSail*®.

Jim Cantrell first worked with The Planetary Society in the late 1980s as a student from Utah State University on our Mars balloon and Mars rover projects. Utah State had a fine aerospace engineering department influenced by Professor Frank Redd, who pioneered small spacecraft development, and Dr. Gilbert Moore, who was incredibly innovative in finding meaningful, exciting, educational projects for young people. Jim and another student, George Powell, went on to successful engineering and business careers after they first helped the society create the first privately funded projects for planetary exploration. Jim became an expert on Mars balloons, and The Planetary Society "lent" him to the French Space Agency to help them develop their proposal for flight on a Russian mission in the 1990s.[11]

Jim went on to work on a number of defense and civil space programs, including remarkable international cooperation ventures on classified defense projects. He also was one of the pioneers of the small spacecraft revolution that grew out of Utah State University. Today he splits his time between aerospace engineering (with his own company) and Trans Am sports car racing. He maintains extraordinary contacts throughout the aerospace industry—from the commercial and academic smallsat developers to the large defense contractors.

A lot of people in the aerospace industry (especially younger ones) were engaged with the new CubeSats and nanosat development.[12] Many industry veterans regarded them as "toys" or as good for education projects but not for anything serious. But it was clear that nanosat spacecraft development was in its infancy and that technology was only going to get better and using it only more practical.

My friend and colleague Tom Svitek is also deeply knowledgeable about CubeSats and small spacecraft. Tom is a brilliant Caltech Ph.D. whom I first knew while he was still a student in his native Czechoslovakia. Tom was interested in solar sails then, and he wrote me a letter (in my capacity as the solar sail development leader at JPL), including a paper he was writing about the subject. We corresponded some and by luck were able to meet during an International Astronautical Congress in Paris in 1982. A couple of years later I arranged to meet him in Prague when I was there for a meeting of the International Halley Watch. Over some good Czech beer in a little pub, Tom told me that he and his family were escaping from behind the Iron Curtain and planning a surreptitious departure out of Czechoslovakia. I agreed to carry some papers of Tom's out of the country, and later, after he and his family made it to Austria (walking out over the mountains while on a supposed holiday), my wife and I made plans to help Tom and his family get to America and get a start here. I carried his college transcript to Caltech, and he was admitted as a Ph.D. student under Bruce Murray.

Tom is extremely focused on building things to fly in space—instruments, subsystems, and, in the case of our *LightSail*®, a complete spacecraft. He started his own company, Stellar Exploration, Inc., in San Luis Obispo, California (chosen because it was a good place to raise his large family and adjacent to an excellent aerospace engineering department at California Polytechnic University). Cal Poly was a leading center for CubeSat development and flight experience.

Tom, Jim, and I roughly estimated that we could build and fly the spacecraft for about $1.8 million. That was more than The Planetary Society could comfortably raise from ordinary member donations. As with *Cosmos 1*, we needed a major donor or sponsor to commit the bulk of the funding before we could engage our members for their support. The serendipitous funding opportunity came in mid-2009 when our development director, Andrea Carroll, mentioned that a society member from Texas had penned a little note on one of our mass mailings saying that he was interested in our solar sail ideas and that he had the means to provide significant funding. After she spoke to him, she was convinced he was sincere. We invited him to meet us in Pasadena. When he asked me (over breakfast), "If I gave you a million dollars, could you do it?" I blurted out, "Yes." I think Andrea was startled at my too-quick answer—but I was sure with that kind of a lead gift we would raise the rest. We shook hands, I got a go-ahead from the

Figure 8. *LightSail®* fully deployed sail. The Planetary Society and Stellar Exploration, Inc.

society's board of directors, and in September 2009, Jim, Tom, and I began the development of *LightSail®*.[13]

Our spacecraft, like *NanoSail-D*, was a three-unit CubeSat with two units devoted to sail storage and one to electronics. As shown in figure 8, with the sail deployed you get the feeling for the very scale of large area and small mass. The CubeSat units are approximately 11 cm on a side, and thus the whole spacecraft (33 × 11 × 11 cm) fits into a piece of carry-on luggage for an air flight. The *LightSail®* design was the first CubeSat spacecraft to be fully functional, with attitude control, radio, cameras, solar and battery power, and an onboard computer for command, control, and data processing. It has twice the sail area as *NanoSail-D*. It was the packaging of all this into CubeSats that convinced me that the nanosat design would lead to interstellar precursors—more on that in chapter 5. Figure 9 shows the packaged spacecraft configuration with the sail folded inside.

We finished the spacecraft development in April 2012,[14] but despite being selected for launch on NASA's Educational Launch of Nanosatelites (ELaNa) program (through a cooperative agreement with NASA's Ames Research Center), no suitable opportunity to launch us to a high-enough orbit had been identified. We were different than almost all other

Figure 9. LightSail® spacecraft. The Planetary Society and Stellar
Exploration, Inc.

CubeSat-based spacecraft in that we needed to launch above 800 kilometers, to be above the Earth's atmosphere so that the spacecraft could fly on
solar pressure without the drag of atmospheric molecules. On May 4, 2012,
after a week of final full-scale end-to-end system tests, mechanical and
electrical testing, and communications system testing, we put the spacecraft
into sealed storage awaiting the availability of a launch opportunity. We
still had work to do—the software was incomplete and an update would
be required before the final launch preparations, and we also were not
fully confident about our radio and planned to investigate alternatives for
a possible replacement. With funds running low, the spacecraft in storage,

and no launch date set, The Planetary Society Board of Directors put the project on hold, where it remained for the next year.

At that point, I stepped down as the project director.[15] The society never got a NASA (ELaNa) launch to a high-enough altitude, but they did get a lower-orbit launch in 2015 for a deployment test with what was our spare spacecraft. With a new team they took the spare spacecraft out of storage and prepared it for the ELaNa launch on an Atlas V rocket. The test was successful; *LightSail's*® sail was deployed in low Earth orbit on June 7, 2015. They also made a cooperative agreement with Georgia Tech to have our prime spacecraft deployed by their *PROX-1* (for proximity) spacecraft, which they are developing for the Air Force. *PROX-1* will be able to observe *LightSail*®, which would be terrific. That launch is scheduled in 2016 on the maiden flight of Falcon 9 Heavy, but the launch date is uncertain. Unfortunately, the altitude of *PROX-1*, and hence of *LightSail*®, will be just a little lower than the desired 800 kilometers—but hopefully high enough for a meaningful flight.

The impetus behind The Planetary Society solar sail effort was to be first—first to initiate the technology that would someday take humanity to the stars and first to actually fly a sail with the force of light. We thought we would succeed when we built the first solar sail spacecraft (*Cosmos 1*), but its launch failed in 2005. We initiated our new attempt with *LightSail*® in 2009, but before we could get it built, the Japanese space agency successfully flew *IKAROS* and won the bragging rights for the first solar sail flight.[16] *IKAROS* earned an entry in the Guinness Book of Records. Metaphorically, the wind had fallen out of our sails. On the positive side, our design had caught the attention of the rapidly enlarging small spacecraft community and was used as the basis for several NASA proposals in their new Edison (small satellite) program and by JPL for interplanetary CubeSat proposals. Our development was being passed on to its descendents.

Several groups are now building solar sail spacecraft and plan to fly soon, within the next few years. Very similar to our *LightSail*® spacecraft is one being developed at the University of Surrey in England. Surrey's is CubeSat based, with a spacecraft called *CubeSail* that is even smaller than *LightSail*® but is intended not for solar sail flying but for operation in low Earth orbit as an atmospheric drag brake to remove satellite debris. This was how *NanoSail-D* was also used.[17] They hope to develop their device commercially to help the orbit debris problem now of concern to satellite operators. Another project with CU Aerospace, an Illinois company working closely with the University of Illinois, is also called *CubeSail* and is also intended to fly in the atmosphere to make measurements. It is somewhat inaccurate to call these spacecraft, admittedly using sails but flying in the atmosphere, "solar sails."

More significant is NASA interest. Very recently, NASA approved two solar sail missions with nanosats based on the *LightSail*® design. One, called *Lunar Flashlight*, has a lunar destination, and one is a *Near-Earth Asteroid (NEA) Scout*. These are significant because they will fly beyond low Earth orbit with scientific objectives (although they are primarily technology test vehicles in the NASA program). Whereas *NanoSail-D* and *LightSail*® are 3U CubeSat designs (three Cubesat units), *Lunar Flashlight* and *NEA Scout* are 6U—basically twice the mass and nearly three times the sail area. But they are still nanosats. With the cancellation of *Sunjammer*, it now appears that one of these two will be NASA's first solar sail.

It seems that the time for sailing the cosmic ocean has finally arrived. The shipyards are busy with the first of the craft, and the technology advances are providing a basis for estimating performance and designs that can achieve the high performance needed in deep space. In the next chapter we will look at the possible performance parameters.

CHAPTER FIVE

Interstellar Precursors and the 100 Year Starship™

Interstellar spaceflight for humanity isn't inevitable—merely imperative.
—SHUTTLE ASTRONAUT MAE JEMISON, PRINCIPAL,
  100 YEAR STARSHIP™

It seemed a bit incongruous when the Defense Advanced Research Projects Agency (DARPA), an agency of the U.S. Department of Defense, initiated a program to consider a 100 Year Starship™ (100YSS)[1] based on the vision of interstellar flight. 100YSS was the brainchild of David Neyland, a program director in DARPA, and Pete Worden, director of the NASA Ames Research Center. They were inspired by a Robert Heinlein story in which entrepreneurs create a Long Range Foundation for investing in future space travel to spur innovative science and technology. DARPA wants to spur such innovation. Heinlein's idea, consistent with his ideology, was that the foundation would be private and entrepreneurial, yielding a return on investment. DARPA's idea was that 100YSS should be self-sufficient after receiving two-year seed funding from the government. But no one (serious) had the idea that real interstellar flight would be self-sufficiently funded or accomplished within 100 years, only that its motivation and development would lead to innovation.

The concept was discussed at an invitation-only strategic planning workshop in January 2011 and at a public symposium in Orlando, Florida, in September 2011. DARPA then solicited competitive proposals to create a non-governmental privately funded organization to pursue the

100 Year Starship™ (with a due date of 11/11/11—a bow to numerology). The emphasis at the workshop and symposium was on human interstellar flight. I was disappointed to hear how little the ideas had changed over all the decades of the space age and how they were still dominated by science fiction authors—brilliant authors, but still writing fiction. It seemed to me that the future of interstellar travel might be less about sending large humans on long voyages and more about sending human surrogates made from advances in robotics, nano-electronics, biology, and information processing. Instead of focusing on propulsion and life support as the key technologies for interstellar flight, perhaps the focus should have been on nanotechnology, biogenetics, and data processing. There were exceptions: J. Craig Venter of DNA-synthesis fame suggested that interstellar flight might involve DNA molecules programmed to interact with the target planet to send information back in ways we can only begin to imagine.[2] Venter elaborates on this and other ideas about the merging of information and biology in his book, *Life at the Speed of Light*.[3] But when I asked the organizers to bring more focus to robotic flights communicating with humans at home, I was outvoted—approximately 25–1. This is still true: 100YSS is devoted to the idea of humans traveling to the stars.

Instead of getting angry, I decided to get even: instead of thinking abstractly about what we can't do, let's think about what can we do now to extend the human presence to the stars. We can't do interstellar flight, but can we create meaningful precursors that can advance both capability and understanding, that is, both technology and science?

In the previous chapter I described an evolution of thinking about solar sail spacecraft—from the enormous nearly one-metric-ton JPL Comet Halley design spanning 15 kilometers in diameter (as the heliogyro), or a half mile by half mile square (in the three-axis stable design), to *LightSail*®, weighing less than 5 kilograms. In this chapter we will see how we can use nano-spacecraft to escape the solar system at very high velocities and thus enable a series of flights that serve as interesting missions in their own right as well as interstellar precursors with milestones on the path to the stars.

Throughout this book I emphasize that human exploration of other star systems will occur by leaving the human at home. Nano-spacecraft will make interstellar starships instead of monster nuclear, antimatter, and warp-drive vehicles. This is how technology is evolving. Whereas human spaceflight technology has advanced hardly at all over the past half century, robotic technologies follow a type of Moore's law (which describes exponentially fast advances in electronics and computational technology). Robotic sensors and instrumentation have rapidly increased capability with less volume, mass, and energy, as opposed to human sensors and manual capability, which have hardly increased capability at all. In this chapter I describe incremental approaches to interstellar flights that are fast, economical, and technically feasible (miracle free), utilizing those increasing advances in robotic technology.

Practical interstellar flight may be advanced with fascinating precursor missions flown in the next several decades to serve as milestones on the way to the stars. A breakthrough can be achieved by propelling nano-spacecraft with large solar sails to fly close to the sun to produce high-acceleration vehicles that escape the solar system at high speeds. Nano-spacecraft today have limited payload and communications capabilities, but that situation is rapidly changing—so much so that interplanetary nano-spacecraft are already being developed for flight within this decade.[4]

In chapter 2 we described the key parameters for a fast solar sail spacecraft escape from the solar system: area, mass, and close passage of the Sun (low perihelion). The area to mass (A/m) ratio (big area, small mass) is proportional to the acceleration imparted to the spacecraft. A big area collects lots of photons, and each photon transfers energy to the spacecraft. The lighter the mass of the spacecraft, the larger will be the resulting acceleration. The close solar flyby permits two things: more power from the Sun and application of maximum acceleration at the most efficient place in the orbit (the perihelion) to increase energy and hence the size of the orbit, extending its aphelion (farthest point from the Sun) beyond the outermost planets. In this way, the aphelion can be extended to infinity—that is, changing the orbit from a closed ellipse to an open hyperbola, extending beyond the solar system.

Closeness to the Sun is limited by the thermal properties of the spacecraft and the sail itself. A sail at 1 AU, the distance of the Earth from the Sun, has only to withstand a temperature of less than 90°C (194°F), but at a distance of 0.25 AU, it has to withstand about 600°C (1112°F).[5] And, as seen later, we will want to go even closer to the Sun. Mylar would degrade, so it is not a suitable material for long-duration and high-radiation exposure, but certain polyimide plastics can handle it. Really advanced materials made out of carbon nanotubes and fibers or with substrates that evaporate leaving only ultrathin aluminum molecules may enable closer solar flybys—although the aluminum in the supporting structure will lose its strength at temperatures greater than 400°C. Thus, the search for an interstellar propulsion system will include finding the best possible materials to make the sail.

Before considering how far and how fast we can go with solar sail propelled nanosats flying close to the Sun, let's examine some interim milestones on the way out of our solar system: waypoints for interstellar solar sails. The distance to the farthest planet[6] in our solar system (Neptune) is about 30 AU. The Kuiper Belt of icy objects like Pluto, many of which are dwarf planets or inactive comets, is located from about 40 AU perhaps to as far as 500 AU from the Sun (although most of them are probably nearer the 40–70 AU distance). In the last decade we have discovered that Kuiper Belt objects (KBOs) are a distinctive class of celestial bodies, not quite planets but different from asteroids, which are generally located within the inner solar system. Asteroids are generally rocky; KBOs are generally icy. If an icy body's orbit gets gravitationally perturbed, it is possible that the KBO can get away and go in toward the Sun, with its ice melting, forming a tail, and thus becoming a comet. When a comet's orbit gets perturbed and becomes smaller, making repeated flybys closer to the Sun, it will burn out, with its ice melting (leaving only a rocky core), and then become an asteroid. There are billions of small KBOs and thousands that are larger—tens to hundreds of kilometers in diameter.[7] The larger KBOs are big enough to have their own gravity affect their internal structure, and they become round. The large KBOs are called "dwarf planets."[8] When Pluto was the only such object known, it was called a planet, but once others were discovered, it became the leading member of the new class of dwarf planets and KBOs. It is the leading member (for both historical and orbit-location reasons) but not the biggest. A KBO named Eris is currently the largest known. In 2015 a spacecraft called *New Horizons* will fly close to Pluto after a nine-year voyage from Earth; after its encounter with Pluto, it will venture into the Kuiper Belt to make new discoveries. Whatever it discovers, there is a lot of territory out there, and there will be a desire for more spacecraft. The Kuiper Belt is the first milestone beyond the planets of our solar system.

The boundary from interplanetary to interstellar space is usually described as the *heliopause*, a broad and irregularly shaped area where the stream of particles emitted by the Sun (called the solar wind) gives way to the stream of particles from other stars, called cosmic radiation. The heliopause is roughly 120–150 AU outward from the Sun (it is irregularly and broadly shaped and does not have the same distance in all directions). *Voyager 1*, after a 36-year journey beginning in 1977, finally passed through the heliopause in 2012—the first emissary from Earth to do so. *Voyager 1* is traveling approximately 3.7 AU/year.[9]

Notice that I called the heliopause the boundary between interplanetary and interstellar space, not the boundary of the solar system. The gravitational influence of the Sun dominates many times the distance of the heliopause, and what we call the boundary of the solar system can be argued. But it is fair to say that *Voyager* has now entered interstellar space, and in that regard it is an interstellar precursor. (See chapter 2 for more discussion of *Voyager* and the interstellar medium.)

I regard *Voyager's* passing into interstellar space as an important human culture milestone—it extends our reach into the galaxy, deeper into the universe. It is also an important scientific milestone because it brings us data from a totally new environment, one where our Sun is less important and where the stuff from other stars dominates. Having one, and in some few years two, spacecraft reach past the heliopause does not mean it is explored or that we can now focus only on the next step beyond. Much remains to be done to understand the structure and nature of that region and how particles and fields behave beyond it. Although *Voyager* has reached a region where the interstellar wind begins to dominate, it is not until beyond 200–300 AU (the exact distance is unknown) that we are in a region of "pure" interstellar wind without solar particles and fields.

The nearest star (besides the Sun), Alpha Centauri, is about 4.3 light-years distant; 1 light-year = 63,000 AU, meaning that Alpha Centauri is 271,000 AU from the Sun! It will take *Voyager* about 17,000 years to go 1 light-year. As far as it has gone, it is still less than 0.005 percent of the distance to the nearest star. Are there any other milestones beyond the heliopause on the way to the stars? We mentioned the 200–300 AU distance of "pure" interstellar wind—probably the result of a barrier or wall of hydrogen molecules created by collisions and interactions of the solar and interstellar wind particles.

Light travels with finite speed and that is affected by gravity. When light passes a massive object, its path is bent. The bending of light from a distant star by the Sun was one of the critical observations made by astronomers that proved that the theory of general relativity correctly described gravitation. In the next chapter we will describe a place where light rays from a distant star bent by the gravity of the Sun will focus—called the solar gravity lens focus. It's actually a line of focal points emanating straight out on the line connecting the star and the Sun beginning at about 550 AU from the Sun. If a spacecraft could stay on this focal line and use the solar gravity lens to magnify observations of that distant star, or its planets, this might be a very valuable and unique way to image worlds around other stars. The solar gravity lens focus could be another waypoint or milestone for interstellar precursors, and in the next chapter, we will consider going there.

Beyond the solar gravity lens focus there is the numeric milestone of 1,000 AU. JPL did a study of a Thousand Astronomical Unit (TAU) mission, not because anything (known) was there but as a distance milestone.

Beyond that is the Oort Cloud—between approximately 5,000–50,000 AU. We don't know exactly because we can't see small things that far away. The existence and properties of the cloud are inferred from dynamical analysis of comet orbits and theoretical models of the solar system. It is presumed to be a spherical shell of small asteroid-like objects, although it is hypothesized that the inner part of the Oort Cloud is more densely populated, with objects distributed primarily in the ecliptic plane.[10] But who knows? This area, fundamental to understanding how the solar system formed, how the orbits of comets and asteroids evolve, and how these orbits affect the history of the planets is unexplored, and it would be a milestone on the way to interstellar flight. The Oort Cloud extends to nearly one light-year (63,000 AU).[11] The outer edge is what we can define as the edge of our solar system. This is about as far as I think is practical to consider interstellar precursors. The Oort Cloud is still only between 2 and 20 percent of the distance to the nearest star, and to get there in less than 50 years would require a speed of at least 100 AU/year (0.16 percent the speed of light). We will see later in this chapter that that is well beyond the state of the art. Maybe it can be done by the end of this century, but even that is a tall order. Beyond that our next step would be to the stars themselves—in another century. The geography of the interstellar medium is illustrated in figure 1 (chapter 1).

For the first of the 100YSS symposia, I joined with two colleagues to help develop and write a paper about the idea of interstellar precursors: proposing milestones outside our solar system that we could reach with nano-spacecraft and solar sails. The two colleagues were Dr. Thomas Heinsheimer, an aerospace consultant with vast experience in spacecraft systems and mission analysis, and Dr. Darren Garber, an astrondynamicist who was just completing a doctoral thesis at the University of Southern California on methods to quickly analyze solar sailing (and other low-thrust) trajectories. Darren was working with us on the *LightSail*® mission, developing a highly sensitive, extremely small accelerometer that could measure the ultra-small sunlight forces on the spacecraft and thus allow precise calculation of the spacecraft's orbit. I was excited about this new development, which his company, Millennium Space Systems, was trying to develop as a commercial product, because I thought that in addition to permitting a direct measurement of the force from sunlight it might also lead to the ability for precise orbit determination.[12] While Darren worked on calculating trajectories for interstellar precursors (results below), he kept his day job at Millennium Space Systems, earned his doctorate in aerospace engineering at USC, started his own consulting company (NXTRAC), and helped his wife have a baby—all in less than a year.

Tom was a colleague whom I first began working with when The Planetary Society started the Mars balloon promotion and development. I referred to the Mars balloon in chapter 4 in connection with Jim Cantrell's work for The Planetary Society. The story of planetary balloons is a very interesting one, and it led the three of us, along with some Russian and French colleagues, to many great adventures around the world in the 1980s and 1990s. But that is a different story, not part of this book.[13] As a balloonist, Tom was more of an adventurer; as an aerospace engineer, he is a specialist in system design with the ability to merge the big picture of mission objectives with innovative technical ideas to accomplish them. As we developed the idea of interstellar precursors, I kept emphasizing that we can only take it so far and that ultimately we will need powerful lasers beaming light over interstellar distances, as discussed in chapter 2. Tom, however, disagrees. He thinks giant lasers are science fiction and that technology will advance so that pure solar sail flight with truly advanced materials will someday take us to the stars. We examined questions of solar and laser sails for interstellar flight as part of a study, Exploring the Interstellar Medium, at the Keck Institute for Space Studies. No magic solution emerged—spacecraft would have to weigh fractions of a kilogram to be propelled to interstellar speeds by solar sails, and laser systems to propel more practical-sized spacecraft would require many tens of gigawatts and be incredibly large. Examples later in this chapter will show that pure solar sailing won't do interstellar flight, and thus I remain convinced that laser sailing is the only true interstellar flight technology. But I admit it will be more than a century before we can even think of the possibility, and who knows what we will learn by then. Garber performed a parametric analysis of spacecraft trajectories to determine how far one can go in a given time, as a function of A/m and perihelion distance. We considered A/m ratios from about 1 m^2/kg (approx. that of the Japanese *IKAROS* spacecraft) to about 1,000 m^2/kg (about 50 percent larger than the value for the Halley Comet rendezvous spacecraft studied by JPL in the 1970s). The Planetary Society *LightSail*® is ~7 m^2/kg.[14] The *Sunjammer* spacecraft, which was supposed to fly in 2017–2018 (chapter 3), was ~22 m^2/kg. Scaling up from *Sunjammer*, we might consider A/m ~100 m^2/kg, e.g., a 100 × 100 m sail with mass ~100 kg. Or, evolving *LightSail*®, we might equivalently consider a 32 × 32 m sail with a 10 kg spacecraft, yielding the same A/m. I think it is possible to extrapolate these ideas to at least an A/m ~1,000 m^2/kg.[15] The JPL design for the Halley Comet rendezvous (chapter 3) had an A/m = 711 m^2/kg. Some Chinese mission analysts recently published a paper proposing a solar sail small spacecraft for an asteroid deflection mission with an area to mass ratio of 550 m^2/kg.

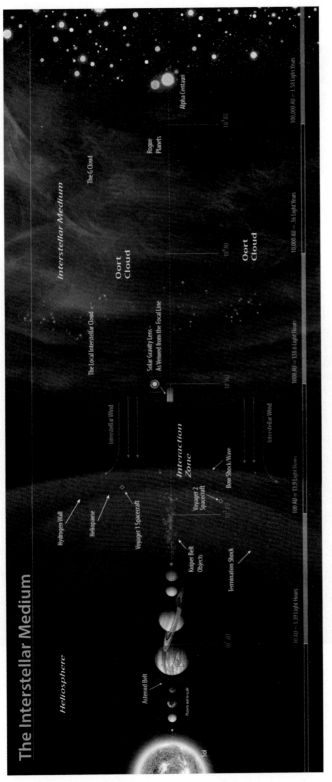

Plate 1. The scale of space exploration. Chuck Carter/Keck Institute for Space Sciences.

Plate 2.
Comet Halley
spacecraft:
heliogyro.
NASA/JPL.

Plate 3.
Comet Halley spacecraft: square sail, flying interplanetary trajectory, view of the back side. NASA/JPL.

Plate 4. Hubble Space Telescope galaxy picture through gravity lens. NASA/ *Hubble Space Telescope.*

Plate 5. Mars beckons: landscape from Mars *Pathfinder*. NASA/JPL.

Plate 6. Mars base with solar sails. Ken Hodges, NASA/JPL.

Plate 7. Terraformed Mars. Michael Carroll.

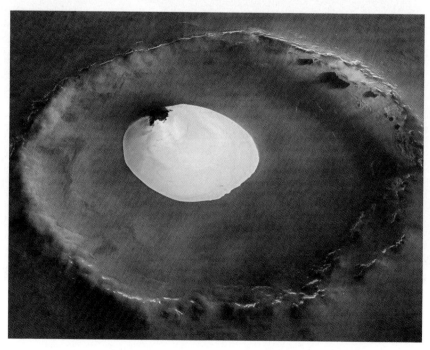

Plate 8. Mars of astrobiology interest; water ice in bottom of crater. © ESA/ DLR/FU Berlin (G. Neukum).

Plate 9. Europa, cracked ice surface showing water evidence. NASA/JPL.

Plate 10. Titan. NASA/JPL/University of Arizona.

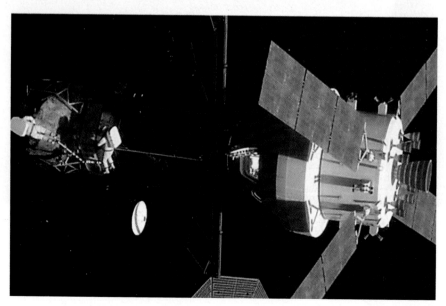

Plate 11. Asteroid visit by crew to retrieved asteroid. NASA.

We considered close perihelion passages from 0.1–0.2 AU. This is really close by current standards—in the 1970s for the Comet Halley mission study, we considered 0.25 AU. But subsequent NASA studies of a solar probe, as well as consideration of carbon nanotube or pure metal sails, suggest that advances in materials will permit closer survivable solar approaches.

Figure 10 shows how far a solar sail spacecraft can go in 50 years as a function of A/m and perihelion distance. Deployment of the solar sail is assumed to be done after launch so that the spacecraft spirals in toward the Sun until the perihelion is reached. The sail is then controlled to raise the aphelion until solar system escape is achieved. The larger the A/m and the smaller the perihelion the faster we can escape the solar system and reach the distances shown on the graph.

With the Jupiter gravity-assist for flying close to the Sun, a spacecraft with A/m = 100 m²/kg and flying to 0.2 AU from the Sun reaches 330 AU in 50 years. Flying to 0.15 AU, it reaches 400 AU, and flying to 0.1 AU, it reaches 500 AU in that same time. For A/m = 400 m²/kg, the distances reached are 550, 650, and 820 AU, respectively. The advantage of lower perihelion became even greater at higher A/m. For those three A/m = 100 m²/kg cases mentioned, the velocity achieved is 6.5, 8.2, and 10.5 AU/year, respectively. For A/m = 400 m²/kg, those velocities are 12, 14, and 17.5 AU/year, respectively. Higher velocities will require bigger sails and lighter spacecraft.

Power will be needed on spacecraft, and so far away from the Sun, it will have to be nuclear. Radioisotope thermoelectric generators (RTGs) can provide moderate amounts of power at high efficiency and without the complexity of a nuclear reactor. They use and convert the heat of radioactive material to electric power. If we use that power in micro-electric propulsion thrusters, we can add to the velocity—at least with larger spacecraft.

Fifty years is an arbitrary time—about the length of a typical professional career. It is longer than we want for practical missions. Not only is our attention span likely to be shorter than 50 years, but at the speed of technology advances, we are likely to invent something that would make the mission obsolete before it reaches its destination. If we want to use knowledge and experience effectively from one mission to design the next, we certainly should try to do missions faster than in 50-year time steps, and below we suggest doing that with increasing technology steps.

It was this idea of a series of ever-increasing interstellar precursors that I presented (on behalf of Garber, Heinsheimer, and myself) at the first 100YSS symposium. As noted earlier, most of the 100YSS discussion focused on larger, human-crewed, classical (fictional) starship missions.

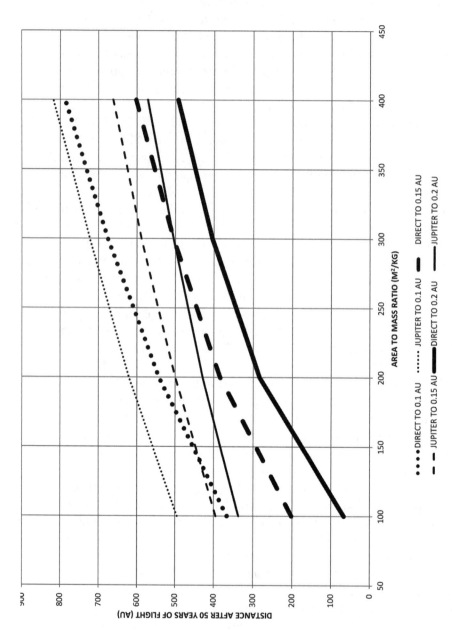

Figure 10. Diagram of distance achieved by a solar sail spacecraft as a function of *A/m* and perihelion distance. © Louis Friedman and Darren Garber.

Not much can be done about that in a 100-year timeframe. For me, the beauty and power of the 100YSS idea was the suggestion that we can do practical things *now* that advance interstellar thinking and flight with projects and initiatives to engage today's generation with the inspiration of reaching other worlds on which there may be life.

The DARPA organizers had the same motivation—they weren't promoting starflight, they were promoting the creation of an organization that would capture its inspiration and innovation. They held a competition inviting organizations to propose how to do that, and instead of picking a traditional engineering or scientific team or one of the interstellar study groups, they picked a new organization—one with emphasis on outreach to a younger generation and one with a global, multidisciplinary outlook. The competition was won by a team put together by the Dorothy Jemison Foundation for Excellence, led by former astronaut, engineer, and physician Dr. Mae Jemison. Mae has picked up the challenge, and in about 97 years (as of this writing), we can evaluate how much closer we are to interstellar flight.

We suggest three example missions that could serve as interstellar precursors that might be possible in this century. They each have interesting scientific targets at successively far-out distance milestones. The flight time for each example assumes a solar sail spacecraft flying at progressively closer distances to the Sun (say from 0.15 AU to 0.05 AU) as the sail materials technology improves from plastics to carbon based. Area to mass ratio is assumed larger for each successive mission, similarly based on the assumption of technology advances. By the middle of this century, we might be able to make sailcraft like those envisioned in earlier paper studies, e.g., 700–900 m^2/kg. For a nanosat of 5 kg, this means a sail with dimensions approximately 60 × 60 m; for a spacecraft of 50 kg, the sail would be approximately 200 × 200 m. These are not impossible values. The larger spacecraft, which we may consider an evolution of the *Sunjammer* architecture, is less constrained on size so as to permit larger instruments and spacecraft subsystems for remote observations and measurements. The smaller spacecraft, an evolution of the *LightSail*® architecture, are the nanosats that will have to be capable of carrying the nano-technology and molecular engineering payloads we imagine for the future.

The goals of the three missions are described in table 2: (1) a mission launched in 2018 to the Kuiper Belt and then going on to the heliopause and the near interstellar medium (ISM); (2) a mission launched in 2025 to go to the solar gravity lens focus and then proceed out to 1,000 AU (along the gravity lens focal line); and (3) a mission launched in 2040 to the Oort

Table 2. Goals of three example missions

	Kuiper Belt / Near ISM	Solar gravity lens focus / Far ISM	Oort Cloud
Distance (AU)	50–150	700 to 1,000	5,000–50,000
~Escape velocity (AU/yr)	10	20	100
Possible mission time	2018–2040	2025–2065 . . .	2040–2090 . . .

Cloud (of course also passing by the intermediate destinations), reaching it by 2090. Ideally, all of these missions would have multiple probes since they all have a vast area to explore—but that will only be economically possible if we really can develop nanosats to carry out low-cost missions. The first two of these destinations, the Kuiper Belt and just beyond the heliopause, are modest extensions of the Voyager and New Horizons missions. They will still enable valuable scientific data to be obtained, both because of new instrumentation and because they will visit different regions than did the earlier missions. The solar gravity lens focus mission (discussed in the next chapter) and Oort Cloud missions would be the true new explorations. The former would also explore the "pure" interstellar medium beyond the hydrogen wall for the first time. This is where all the effects of the solar wind are gone (approximately beyond 200–300 AU). Such missions will also be precursors to interstellar flight both as technical milestones (going farther and faster) and as technology steps—advancing the sail and spacecraft technologies that someday will take us to the stars.

The mission dates in table 2 are only suggestions as to what might be realizable for launch and for reaching its milestone destination. The ". . ." is meant to imply that scientific measurements at the destination will continue for years afterward as the spacecraft travels farther. The example dates show a sequence of three interstellar precursors of ever-increasing complexity and achievement as well as of progress toward interstellar travel. Looking at the scenario created by these dates, we have an impression of looking at a glass both half-full and half-empty.

On the one hand, we have three exciting new missions of exploration moving into interstellar space with the potential for new scientific discoveries and technological achievements. And we can begin on this right *now*, with an ambitious goal of launching the first one in 2018, the second just seven years later, and the third within fifteen years after that. Each would have spacecraft with development based on their predecessors, and all three would yield results in this century. Indeed, a half-full glass, at least.

On the other hand, even with this aggressive pace of mission approval (short-circuiting what now is more commonly a decade or two for mission concepts to be approved) and technological development leading to a fully instrumented, robust, capable small spacecraft with A/m ~800 m²/kg as cited above well beyond today's best value of ~8 m²/kg, it still will take us almost the whole century to realize these missions. And by the year 2100 we will have barely reached about 6,000 AU—not even 1/10 of the way out of the solar system. That glass is half-empty. We might double this performance with the invention of some nuclear propulsion boost that can be packaged with mini-thrusters into small spacecraft, but that will still be far short of interstellar goals.

These are interstellar precursors—exploring the interstellar medium, making measurements of interstellar targets, and advancing the only known technology that can take us to the stars. But, they need something more if they are going to lead to what actually takes us to the stars. They need external energy—laser beams (or other electromagnetic radiation beaming) replacing the solar energy, fired over interstellar distances. The above examples show us that beamed energy will be needed.

Reaching a 1 light-year milestone, let alone the 4.3 light-years distance to Alpha Centauri, is beyond the reach of pure solar sails (without laser or other microwave beaming), based on our current understanding. I calculated with all the optimistic assumptions that I could think of that it would take a sail with A/m ~one million square meters per kilogram (e.g., one kilometer square per kilogram) traveling 0.02 AU from the center of the Sun (that is 3.2 solar radii from the surface) about 100 years to reach the nearest star.[16] Laser beaming from a large array in space seems more practical to me. That is actually a technology we know—albeit still quite an engineering challenge to solve. As mentioned in chapter 2, Professor Phil Lubin is conducting research on laser propulsion for interstellar flight, and his latest estimates are that it would take about a 70 gigawatt (70 million kilowatts) laser power from a more than 10 × 10 km solar array in space to propel a 100 kg spacecraft with a ultrathin-film sail about 1 km in size to a velocity about 3 percent the speed of light.[17] This is beyond our current capability, and that's why we say we are still more than a century from practical interstellar flight. However, this is still likely to be achieved before sending humans anywhere beyond Mars.

It may well be another century before such large laser power sources can be readied, but low-cost and low-power beamed energy experiments can begin much sooner to develop this ultimate technology for interstellar flight. We might even be able to use nearer-term, more modest laser beaming to accomplish our interstellar precursor goals, developing both the technology and advancing the science.

Meanwhile, the interstellar precursor missions could be started now—the technology and technical capability exist, and starting on the interstellar path would provide a galvanizing force for the next generation of space professionals—reaching beyond the limits of this generation that has gotten through, but not beyond, the solar system. There is a lot to be developed and tested—nanosat reliability and robustness, power and communications far from the Sun, the materials properties of the sail getting close to the Sun, attitude control, etc. The beauty of nanosats is that we can develop and test these things in simple flights at a comparatively low cost. University research projects could build nanosats to test sail materials and other candidate interstellar precursor spacecraft components and devices. This fits the *100 Year Starship*™ idea to begin on the interstellar path now even if it will take a century or more to actually initiate an interstellar mission. By the time we do those three interesting missions given in table 2 (or ones like them), technology and knowledge will have greatly increased, leading to new ideas and capabilities to tackle the interstellar question. In the next chapter we will consider in more detail the particular milestone of the solar gravity lens focus with a mission that might help define the objectives and the means for the first interstellar flight.

Whether it is the nanosat or larger microsat spacecraft that will actually evolve into interstellar capability remains to be determined. Sail area can be expected to grow from the present in steps—each with lighter weight and stronger booms, probably using carbon fiber or nanotube material in the latter stages. Mass will continue to decrease below the nano-spacecraft level to one-kg, pico-satellite size. Such satellites are already being studied and designed at the Aerospace Corporation, but what capability they will have for long-life, deep space travel remains to be determined. For the *Sunjammer*-derived spacecraft, similar technology advances can be predicted; after all, *Sunjammer* was just to be a test of current technology, severely constrained in cost and ambition. We submit that although speculative these advances are realistic and that with them a doubling of capability every decade for reaching farther and going faster beyond our solar system can be achieved. This might suggest a type of Moore's law for spaceflight, driven not by the marketplace but by a vision to understand our place in the universe.

This chapter has concentrated on science and trajectories—how to get to interesting destinations fast. Much more needs to be done to study the engineering development of solar sail nano- or micro-spacecraft as interstellar precursors. We have to communicate with our spacecraft, and we need a small amount of on-board power for the low mass instrumentation. Spacecraft technology is advancing rapidly, in many cases using dual-use technology from the commercial sector. Advances in radio and optical communications, computing and data storage, miniature radioisotope generators, lightweight materials, and perhaps nanobots using chemical and biological processes in sensors will enable long-life interstellar precursors. Artificial intelligence and other advanced information processing will connect our brains to the spacecraft and the sensors. These subjects are addressed in the next chapter where we consider a specific, possible mission. Even without knowing exactly how, we can bet on these kinds of technology improvements as the most likely ways to extend human presence beyond the solar system.

By the time we advance human flight beyond Mars (centuries from now), we'll be very well served by the virtual exploration created from the robotic craft. But, the lure of interstellar flight should not be underestimated. Whether or not extraterrestrial life is found in our solar system, it will not satisfy the human drive to understand our place in the universe. The extraordinary variety of planets has whetted the desire to find and ultimately explore habitable worlds—worlds that may harbor life independently evolved from Earth's or life from Earth, setting up a galactic species. The magnitude of space is daunting, and interstellar flight has seemed as far from our generation as was aerodynamic flight from DaVinci's. Thinking small, with nano-spacecraft as part of our virtual human brains, reduces that magnitude and brings our access to worlds in other solar systems much closer.

CHAPTER SIX

A Mission to the Solar Gravitational Lens Focus

Time and space and gravitation have no separate existence from matter.
—ALBERT EINSTEIN'S ONE-SENTENCE DESCRIPTION OF
GENERAL RELATIVITY

In the previous chapter we defined a series of missions that would escape our solar system and could, in a practical way, be accomplished within reasonable times in this century. These could serve as interstellar precursors, that is, as technical and scientific milestones that advance both the technology and scientific understanding for true interstellar flight. Perhaps the most interesting of these would be a mission to the solar gravitational lens focus—most interesting because it is the most unknown. As we will see in what follows, such a mission serves as a very challenging goal—not just for the engineering accomplishment but also for what might be learnt scientifically.

First let's consider what and where the solar gravitational lens focus is. Gravity influences all matter—even matter that has no mass itself, such as pure light. Einstein's theory of general relativity describes gravity as a curvature of space and predicted that when light passes by an object of large mass its path would bend in that object's gravitational field. It was the famous bending of light from a distant star by the Sun, observed in 1919 by the British astronomer Arthur Eddington, that provided the first empirical proof of general relativity—and made Einstein a household name worldwide. Light, say from a distant star or galaxy, passing close to

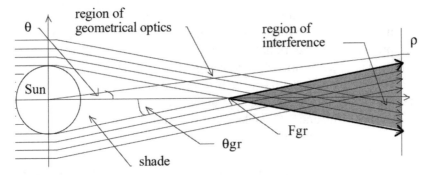

Figure 11. Geometry of the solar gravity lens. Slava Turyshev.

the Sun is bent—more when the light passes near the Sun and less when it passes farther away. Light rays passing on different sides around the Sun converge at a focus, which, because of the differential bending of the light rays at different distances from the Sun, is a series of points forming a line coincident with the line from the star through the center of the Sun. This is seen in figure 11 where we have many points on the line where the light from a point source lined up behind the Sun will focus—a whole line of focal points. When you work out the math of the angle that the light is bent, you find that the first point (nearest to the Sun) is 547 AU along that line.

This would exactly be the case if the Sun were a perfect static solid ball—but it isn't. The Sun has an active and dynamic atmosphere, the solar corona, with many particles that affect the beam of light passing close to the Sun. The light is not bent smoothly, and the actual minimum solar gravity lens focal distance is more like 700 AU. We set this as our mission goal—and in fact define the focal line from 700 AU to 1,000 AU as the mission objective for a solar gravity lens focus (SGLF) mission.[1] A solar system escape trajectory for the SGLF mission could be timed to fly up a focal line (away from the Sun) corresponding to a pre-selected star identified as a scientifically interesting extrasolar planetary target—for example, an identified habitable planet. It could be the target for the first interstellar mission.

Just reaching the solar gravitational lens focus in a reasonable amount of time is challenging enough. A solar system–escape speed of 15 AU/year would reach 700 AU in 50 years and 1,000 AU in 67 years. From figure 10 in chapter 5 we see that this requires a solar sail spacecraft with an area/mass ratio of about 275 m²/kg. For a nano-spacecraft (e.g., 9 kg), this requires a sail area of about 50 × 50 m. For a larger "smallsat" (upper limit of microsats) of 100 kg, the sail area would be ~166 × 166 m.

Current solar sails are using film (plastic substrates on which the reflective metal is deposited) of five-micron (millionth of a meter) thickness. We can probably extend current technology to one-micron-thick sails. A sail of this thickness has an areal density of about 0.64 grams/m², and thus a 50 × 50 m sail would weigh 1.6 kg, and a 166 × 166 m sail would weigh 18 kg. This tells us (qualitatively) that in principle we can design a solar sail spacecraft to reach the gravitational lens focus. In the case of a nano-spacecraft, the sail area is not much larger than that which was being built by L'Garde Corporation for NASA (albeit on a much heavier spacecraft), and we would have approximately 8 kg for the functioning spacecraft, approximately the same as JPL planned for *Lunar Flashlight / NEA Scout*. In the case of the more conventional smallsat, we would have approximately 80 kg for the functioning of the spacecraft. The sail areas are large, but not outlandish—after all, we were considering a 640,000 m² sail for the Comet Halley mission at JPL in the 1970s.

So, we can say that the lens focus is a practical goal, but what can we do there? Maybe we can use the lens as a natural telescope (or signal amplifier) to observe a promising habitable world around a distant star. If we could do that, we might find enough information to make it the prime candidate for an interstellar probe. Identification of such a promising world would come first from Earth-based observations. The focal line defined by the line between the distant star and our Sun would be the target for our SGLF mission. However, the gravity lens is complicated, resulting in not just a single image at a focal point but a ring (known as the Einstein ring) along the focal line. The focus is not a point, but a smeared image in that ring. An example of an image in the gravitational lens is shown in figure 13, taken by the *Hubble Space Telescope*, with a galaxy creating the gravity lens containing the image of a galaxy behind it.

My colleague Slava Turyshev, together with John D. Anderson, analyzed the increase in brightness of a distant star system whose light is bent by the Sun to its focus beyond 550 AU. The gain in brightness is over one million! A one-meter telescope at the focus yields a collecting area larger than one billion square kilometers, i.e., equivalent to a telescope with a diameter of 37 kilometers! The trouble, of course, is that the starlight will be magnified as much as its comparatively dim little habitable Earth-like planet and picking the planet out of that bright light will be difficult. That is routinely done with sunshades, but it remains to be investigated how it could be done after magnification by the solar gravity lens. One idea is to fly along the focal line outward from 550 AU (e.g., to 1,000 AU or farther). The angular resolution of the observed image changes as we fly out of the focal line (inversely

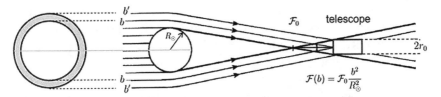

Figure 12. The image using a gravity lens focus. Slava Turyshev.

Figure 13. Hubble Space Telescope galaxy picture through gravity lens. NASA/ Hubble Space Telescope.

proportional to the square root of the distance). It is small enough to resolve large features on the planet's surface, to see an Earth-like planet tens of light-years from the Sun with a resolution equal to that of Earth-orbiting satellites seeing features on Earth. If we can fly right down the focal line, and keep pointing accurately, we will have time for a high degree of signal processing, which can be used to extract the little planet's photons from its star's. Sun shades, like that being planned for use on the *James Webb Space Telescope*, might also permit planet observations—although at the scale of our observations, it is likely they will be electronic.

So, all we have to do is hold the spacecraft steady, fly on a nearly perfect straight line, gather the image in the ring, and analyze that image, getting rid of our own bright Sun, the bright extrasolar star, the smearing effects of the solar corona, the distortion of the Einstein ring, and then presto—we have the image of the exoplanet on which we can do a spectral analysis to determine its atmospheric composition and look for signs of life.

Did I say "presto"? The process to produce that exoplanet image as described above is, unfortunately, still a big bit of magic that will have to be invented. Fortunately, however, Earth-based astronomers already use gravity lens imaging for distant galaxies lining up behind a star relative to Earth, so if we can capture the data in enough detail with onboard instruments and send it back to Earth over the long communications distance, then we can leave it to Earth-based astronomers and computers to produce that image and extract the other data. Dealing with the Sun's corona will likely require a spacecraft coronagraph, something spacecraft engineers are now used to implementing on missions to look for exoplanets.

Light is not the only signal that reaches across the universe. Astronomers also use other electromagnetic frequencies. For detection of extraterrestrial *intelligent* life, "listening" to radio signals has been tried for 50 years. It was once thought that there must be many intelligent species in the universe that would have developed radio communications and be using it to signal their presence (or have it accidentally detected). However, we have yet to hear anything, and the prevalence of radio communication, or even of intelligent species (whatever intelligence means in this context), is much more questioned. My own view is that there seems to be a lot of hints that life will start if the conditions are at all reasonable—after all, it started and took hold on Earth when the planet was still an infant planet in a hostile solar system environment. But the evolution of life on Earth seemed to take a very long time in comparison to its origination and single-cell stability. And it seems to have been influenced by several special, perhaps random, events (like species extinctions from extraterrestrial impacts). So maybe evolution is hard and evolution to complex, intelligent species (again, putting aside what "intelligence" means) even harder. In short, habitable worlds may be prevalent, but intelligent species may not be. Search for extraterrestrial intelligence (SETI) enthusiasts have not given up: some are switching to optical searching, and some radio proponents keep improving their electronics and radio telescopes, arguing that so little of the search-space has been adequately covered.

Some things have changed to diminish SETI's intellectual standing. First, we have 50 years of no results. Also, the idea of special frequencies defining likely communication channels has become weaker—primarily due to interstellar scintillation of radio signals preventing us from choosing a likely frequency without knowing its distance traveled. Finally, there is the hypothesis shift away from "intelligence" being what we should look for to "habitability" of newly discovered extrasolar planets being a prime observation goal. About one thousand extrasolar planets have been discovered, and it seems pretty clear that thousands more will be within the next few years. Star systems having planets may be the rule, not the exception, and the huge diversity of them indicates that habitable planets are prevalent—maybe even to the point of there being an average of one per solar system around other stars. What we don't know is the connection between habitability and life, i.e., how easily life forms if the conditions are right. This is an exciting question to ask, and one we are on the cusp of answering. But, to me at least, all these positive discoveries about numerous exoplanets and likely numerous habitable ones are negative evidence for SETI. Given the billions of years over which planetary formation has been going on among the tens of billions of stars in just our galaxy, if so many habitable planets have been formed and not one of them has an intelligence beaming recognizable signals to us (or otherwise being noticed over the past half century), then communicating intelligence must be rare, at best. The Fermi paradox[2] (asking "where are they?") is now more likely to be the Fermi theorem—with the proof being the number of discovered intelligent species divided by the number of habitable planets in our galaxy.

An aside here to call attention to my SETI skepticism: as executive director of The Planetary Society, I not only helped create privately funded SETI programs when government support was withheld but also championed their virtues in active fundraising and advocacy campaigns. Perhaps I should admit to a little bit of disingenuousness in doing that, but there were and are four strong factors that justify creating and conducting SETI even while being skeptical about its possibilities for real discovery. First (and most important), I might be wrong. If we don't look, we won't know. Second, SETI was comparatively low-cost and the observing program was scientifically sound. More than just "sound," it attracted brilliant scientific minds— such as Phillip Morrison, Barney Oliver, Carl Sagan, Frank Drake, Jill Tarter, Paul Horowitz, and many others[3]—and they in turn engaged graduate students and young researchers who do outstanding work. Third, it was conducted with terrific engineering advances in radio astronomy and computing. Even without the intelligent species result, those advances were worth the cost. Fourth, it is inspiring and educational. Looking into the nature of life in the universe and the possibility of either its ubiquitousness or uniqueness provides an enormous learning opportunity and motivates young and old alike to learn more about the science of the universe. For those four reasons I had no intellectual qualms advocating for SETI even while skeptical about extraterrestrial intelligence beaming signals at us. I am proud of the role The Planetary Society played in creating a bridge to rescue SETI when government support stopped in the United States and in other countries and then in helping pave the way for the full-time dedicated SETI program of the SETI Institute in the 1990s. The society pioneered private SETI funding as well as private funding for the detection of extrasolar planets and the discoveries of near-Earth objects (asteroids and comets) long before these subjects were popular and long before private funding became a commercial activity.

When considering missions to the SGLF, we must mention the detailed work of another colleague, Claudio Maccone, an Italian engineer and astrodynamicist who has long championed SETI efforts, including an idea to build a radio telescope on the Moon to listen without Earthly signal interferences. Maccone is an enthusiast for using the solar gravity lens to focus putative radio signals from an intelligent species and capture them on a SGLF mission that he calls FOCAL. He has even written a book all about using the SGLF in radio astronomy.[4]

A radio signal from a distant star (let alone one that shows its planets) would be much less powerful than the light signal. In addition, the gain of the solar gravity lens is several orders of magnitude less than at radio frequencies. But, if we can use the solar sail as an antenna and thus have an effective antenna with diameter ~70 m, then the antenna gain would effectively double the gain of the SGLF. Using assumptions made in SETI, it would be a great place to hear an extraterrestrial-beamed radio signal—if there were any such thing. Despite my skepticism about that, I would very much want to instrument my spacecraft with a radio signal receiver if I could do so within mission constraints since I will be at the right place with my solar sail antenna, and as noted, if we don't listen, we can't hear.[5] For other reasons (communication with Earth), we will want to study the practicality of using the solar sail as an antenna.

In the previous chapter we specified two examples of spacecraft that would enable reaching the SGLF in less than 50 years, i.e., 15 AU/year to reach 750 AU:

* the nanosat with a 50 × 50 m sail propelling a 8 kg spacecraft
* the smallsat with a 166 × 166 m sail and ~100 kg spacecraft

I believe such spacecraft are reasonable, although as noted above, 50 years is still a long time. Lighter spacecraft and larger sails are more desirable, as will be getting a lower perihelion. Maybe, also, we can use a nuclear power electric propulsion boost as mentioned in the previous chapter and reduce the flight time to nearer 25 years. In addition, designing the spacecraft for such missions will be very challenging extrapolations of those being designed today in NASA programs and, in the case of nanosats, by the CubeSat community. The challenge is not merely building the light-weight solar sail spacecraft but also instrumenting it with required onboard capabilities for science measurements, communications, data processing, command, control, stability, and navigation. Together with Ed Stone[6] of Caltech and Leon Alkalai at JPL, I co-led a study, Science and Enabling Technologies to Explore the Interstellar Medium, at the Keck Institute for Space Studies, considering the questions of what is practical for space missions to hundreds of AU. Figure 1 in chapter 1, showing the scale of the space exploration, is a product of that study.

The rest of this chapter will examine what we have to do to create a working SGLF interstellar precursor spacecraft—looking at the spacecraft systems to identify what is practical in the next decade or two to launch such a mission. We will primarily discuss doing this within the nanosat design because (a) if we can do it there, we can do it anywhere, that is, for a larger spacecraft, and (b) intuitively I believe that the super large-size solar sails (hundreds of meters to kilometers in diameter) will prove less practical, with more "devils" in their details than ultra-miniaturization of spacecraft, which seems to be the way of evolution. Certainly they should be more affordable and hence have a better chance for real project implementation. I discuss below the main spacecraft considerations: power, communications, attitude control, and stability.

Power

Working beyond the solar system with small spacecraft presents a power challenge—probably the biggest one for the ultimate spacecraft design. Solar power is negligible, and generating nuclear power has been inefficient and heavy. Deep space missions have used nuclear power sources, where a radioisotope's[7] heat (from decay of the radioactive source) is thermo-electrically converted to electrical power in a radioisotope thermoelectric generator (RTG). RTGs were used on the first Mars landers (*Viking*) as well as on the *Pioneer*, *Voyager*, *Galileo*, *Cassini*, and *New Horizons* space-craft to the Outer Planets. They fell out of favor because of the politics of nuclear energy, but as both the need and their safety have become better understood, they are still being planned for future missions. RTGs are not to be confused with nuclear reactors, which require large controlled chain reactions to provide larger amounts of power.

Spacecraft RTGs to date have weighed several tens of kilograms for 100–300 watts. Small radioisotope power systems (nuclear batteries) might, however, in the near future provide about 20 watts of electricity per kilogram (an improvement by a factor of five over current systems).[8] There have been some hints that the U.S. Department of Defense has classified studies of such nuclear batteries, perhaps no larger than one kilogram.

It should be possible for nanosats to carry a 1 kg RTG, and 20 watts of electricity might be sufficient. Much lighter power sources (tens of grams) are radioisotope heating units (RHUs), which use only the thermal heat of small radioactive sources. They are very small, about the size of a quarter, and generate only milliwatts of power, e.g., 200 mW. One of these certainly would not be enough to meet computer, control, communication, and instrument requirements, even for a nanosat. But could 10 of them be distributed over a large spacecraft area (even "pasted" on the lightsail) and harnessed to provide several watts of power for a fraction of a kilogram? Perhaps RHUs could be used to trickle charge batteries on the spacecraft, which would provide sufficient peak power for the relatively rare operations periods on an interstellar precursor.

As hard as it is to think about generating and operating on minute amounts of power hundreds and thousands of AU from the Sun, it is easier than imagining powering human life support over that distance and for the lengths of time involved. The difficulty of power engineering is actually another argument strengthening the thesis of this book, that human spaceflight will extend beyond Mars only without humans on board.

Communications

Receiving a radio signal over 700 AU requires a high frequency, a large antenna, and significant spacecraft power. The gain of the communications systems depends on these three parameters and, of course, on the distance between the spacecraft and the receiving station.

Ralph McNutt and his colleagues have done one of the few spacecraft design analyses for extrasolar system missions in a NASA Institute for Advanced Concepts (NIAC) study, called the Innovative Interstellar Explorer. They found that with a Ka-band radio system (approx. 30–40 GHz — 10 to 100 times that of UHF-band), a 10-watt spacecraft transmitter, a 14 m spacecraft antenna and a 34 m ground antenna, they can achieve a data rate of 20 bits/second at 1,000 AU. 20 bits/sec is a very low data rate, but it would be just fine for us; after all, we have lots of seconds to accumulate the data on the ground. The limiting factor for communications is the power of the transmitter, and perhaps by using the whole sail area (or even half of it) as an antenna, we can lower the power requirements. If the antenna is 5 times larger in linear dimension, then the power can be 25 times less, i.e., less than one-half watt. This might be feasible even in a nanosat.

Future communications systems can use optical wavelengths (with lasers and telescopes instead of radio transmitters and antennas). The laser allows the communications beam to be focused, and thus much less power is required than for the less focused radio beam. A 1 m optical telescope combined with a 10-watt laser transponder can transmit 500 bps from 1,000 AU to a 4 m Earth-orbit-based telescope (from the same McNutt NIAC study). However, the pointing and stability requirements for the laser are pretty tight—about 0.7 μradians (approximately 40 millionths of a degree). A JPL team under Jeffrey Nosanov conducted a different NIAC study of the technology requirements for missions out to about 100 AU. They are considering a laser link for optical communications.

I am not sure whether optical communications will win out for interstellar precursors, although I recognize that the spacecraft communications advanced technology engineers really like it. If we can succeed in making our sail a long-life antenna for deep space travel, then radio communications still may prove feasible. Or maybe someone will invent something really new and different. Phil Lubin, the laser physicist at UC Santa Barbara whom I mentioned in connection with laser beaming for solar sail propulsion, has also suggested how the laser beam might be used for communications. A colleague of mine, Dr. Darren Garber (a co-author on our first Interstellar Precursors paper[9]), has suggested stellar occultations to create a Morse code communications system. He wants to use the lineup of distant stars and the sail spacecraft as seen from Earth to "blink" a code of on-off with occultations of the stars caused by the spacecraft. He has called this idea to encode mission data on occulted starlight for deep space communications the Multi-spectral Occultation Relay System Experiment (MORSE). Using Morse code for advanced technology deep space communications would be the height of irony.

But in any of these schemes, the speed of light is a limit—a fundamental limit imposed by the known laws of physics. That limit means that communications will be slow: a signal sent from 1,000 AU will take some 5.5 days to reach Earth, as will, of course, a command sent by us to the spacecraft. There are two theoretical concepts that might overcome the fundamental speed limit. The first is harnessing quantum entanglement (mentioned in chapter 2), and the second suggests communicating through wormholes. Both have a theoretical basis in physics but no basis (at least not yet) for technology. Quantum entanglement refers to the peculiarity of the quantum state of subatomic particles being indeterminate until they are observed. If two such particles, separated over vast distances, have the same quantum state, then in theory observing the one here at home will determine the state of the remote one—maybe meaning we can use that quantum bit (qubit) for encoding information. "Maybe" is arguable. The existence of wormholes is a theoretical conclusion of the multidimensional nature of space-time in general relativity, even speculated upon by Einstein. Making a wormhole even big enough to carry information whose apparent speed would be faster than light due to relativistic clocks is only a science fiction idea. It's an exciting idea, but not a known practical one.

It is likely that we won't overcome the speed limit of light (63,000 AU/year.) But compared to our mission times (many tens of years), the extra 5.5 days of communications time is minor. The problem is not really communications time; it is that the spacecraft will have to operate autonomously and that any communications from or to it will be slow and of very low data rate.

Attitude Control

Since the communications time (time for the radio signal from the spacecraft to Earth) is more than 90 hours for the case of a SGLF mission, the control of the spacecraft will have to be autonomous, at least over periods of months. It will be possible to update the spacecraft computer but only at widely spaced time intervals. In addition to the long round-trip communication time, the economics of monitoring a mission lasting 25–50 years demand minimization of intensive ground support. Attitude control should be achieved with little, if any, propellants or consumables. Sailing does this, but lightsailing in the outer solar system and beyond is essentially impossible. Far from the Sun the sail will no longer be useful for energy, and some suggest we may drop the solar sail once we get past Jupiter. However, as noted, we may be able to use the sail for communications, and we might also be able to take advantage of the sail by incorporating the spacecraft of the future into the sail itself—essentially printing our spacecraft on the sail. The Japanese IKAROS mission has cleverly pointed the way. They used very low-power LEDs imbedded in the sail that could be turned on or off to change the reflectivity in different parts of the sail for attitude control and silicon solar power cells imbedded in the sail for supplying power. Beyond the reach of the Sun's power, perhaps we can imbed the little radioisotope heating units with the LEDs in the sail. For control we can try to think of magnetic approaches interacting with the very weak electric fields of deep space. The *Voyager 1* spacecraft reached 120 AU in 2013 and sensed an increase in the charged particle environment from galactic cosmic radiation. Maybe we can use that field interacting with our large reflective sheet and imbedded power source to control the spacecraft hundreds, and even thousands, of AU from Earth.

Another approach is to spin the sail. The spin will create a stable spacecraft pointed in one direction without requiring frequent control. Spinning designs are very good for fields and particles measuring spacecraft, and they should also be quite adequate for measurements that require only pointing in one or a few fixed directions. This will be the case with a SGLF mission.

Stability

Stability of the spacecraft is important. While it is cruising it may not matter how much control or how much stability we have, but during periods of communication with Earth, it certainly will. If we reach the SGLF and try to make observations of an extrasolar planet there, control will have to be very precise and stable. A spin-stabilized spacecraft design may permit indefinite stability, which when coupled with magnetic or kinetic devices give the ability to control the spacecraft's pointing direction. A three-axis spacecraft (one that has its stability controlled actively in roll, pitch, and yaw) can have a closed-loop system that senses the motions of the spacecraft from perturbations and instantly corrects for them—but correcting for them requires fuel, and we do not want to rely on consumable fuel for long missions in interstellar space. Furthermore, while cruising on these missions, even if pointing at something, we are unlikely to require the ability to move instantly, or even fast. Spin stabilization should work just fine. The *Pioneer* spacecraft (*Pioneer 10* and *Pioneer 11*, our earliest Outer Planets spacecraft) were spin stabilized and have held their stability for decades without propulsion. Not having to actively provide propulsion for control and stability has another advantage—long times of flight on ballistic paths provide a pure opportunity to measure gravitational influences. Gravity from a Planet X (an undiscovered planet or other celestial body), mass distributions from small objects, relativistic effects, and possible anomalies from unknown or otherwise unmeasured physical forces can be detected in the tracking data. The *Pioneer* spacecraft were so unperturbed that tracking data of their trajectory through the solar system revealed an anomaly that at first some thought could be the discovery of a new force in physics. It was not observed on other spacecraft because these were the only long-flight-time spinning spacecraft without active attitude control to be so precisely tracked. When I was executive director of The Planetary Society, we raised money from our members to help fund the research to uncover the mystery of the *Pioneer* anomaly. Data taken over decades of time had to be recovered—some of it stored on out-of-date media and recovered with out-of date machines (like going back to capture concert recordings on wax records and transferring them from record player format to digital recordings to play on iTunes). It turned out that the anomaly wasn't new physics at all. It was due to the spacecraft itself—an asymmetry in the radiation of heat providing a minute but measurable net force. But what the research discovered also was that such a minute force would only be recognizable on a spin-stabilized spacecraft because it did not have propulsive bursts to orient and keep the spacecraft stable. Ultimately, spin stabilization for planetary spacecraft fell out of favor because of the high degree of maneuverability and pointing required for planetary imaging. But for flying in empty space for long distances they are stable.

Gravity measurements and analyzing long arcs of tracking data will be a prime scientific goal of interstellar precursors. A mission making its way from the nearby Kuiper Belt out toward the Oort Cloud should be able to measure both gravitational perturbations (like those cited above) and non-gravitational perturbations from the interstellar medium: particles hitting the spacecraft and minutely deflecting its path.

Can we build a spacecraft with A/m ~275 m^2/kg flying 0.15 AU from the Sun and then exiting the solar system at ~15 AU/year with capabilities to make measurements, process data, and communicate information to Earth over 50–67 years while it flies from 750–1,000 AU? I believe that either of our example spacecraft is possible: the approximately 8 kg nanosat with a 2 kg, 50 × 50 m, 1-micron solar sail, and the 80 kg spacecraft with a 16 kg, 166 × 166 m, 1-micron solar sail. The difficult question is whether the instruments, specifically the camera(s), can be engineered for such spacecraft. In fact, I think much higher A/m will be achievable—if the technology missions now underway are successful. The nanosat poses the bigger challenge: instrumenting it with high capability requires a lot, but because of their lower cost, perhaps we can be innovative in using many of them to cope with some of the engineering and mission design problems. We certainly want lots of them to cover the large space to be explored.

A mission to a SGLF of a particular object is of course able to exploit the gravity lens for only that object. Whatever object is chosen, presumably an interesting extrasolar planet about which hints of habitability have been discovered, it will be done prior to the mission's launch and based on Earth-based or at least near-Earth measurements. No matter how interesting, there will be huge uncertainty about choosing the extrasolar planetary target, and there will probably be at least dozens of candidates that we will really want to investigate. Dozens of different extrasolar planets imply dozens of different solar gravity lens focal lines—and hence dozens of SGLF missions that we would like to perform. With reference to table 2 in chapter 5, this is also true for the earlier interstellar precursor: the Kuiper Belt / heliosphere explorer. Current studies suggest that at least a dozen spacecraft are desired to map the heliosphere and explore different Kuiper Belt objects. If this is to be done, then it must be done with low-cost spacecraft: tens of million dollars each, not hundreds of millions. The missions are all risky and their scientific results are all speculative—that is what makes them exciting explorers. The risk and speculation means that costs must be relatively low, and redundancy and robustness (in the form of backups and additional spacecraft) must be large. The Planetary Society's $4 million *LightSail*® and NASA's $15 million *Lunar Flashlight / NEA Scout* are great beginnings—but instrumenting and configuring future spacecraft to work for tens of years and hundreds of AU from Earth (not to mention the 22-million-mile flyby of the Sun) will take a lot. This is why I am betting on the nanosats—a few kilograms (no matter how sophisticated) can only cost so much.

In the next chapter we will explore some ideas about how the human presence will be extended into interstellar space in nano-spacecraft and why this means that human spaceflight of the future will take place without humans on the flight.

CHAPTER SEVEN

Extending Human Presence, Not Humans

The real voyage of discovery consists not in seeking new landscapes but in having new eyes.
—MARCEL PROUST

In the previous chapters we have suggested that missions beyond the solar system can be accomplished with nano-spacecraft with ever-increasing capabilities. These robotic spacecraft will satisfy our science goals as they explore deeper and deeper in space. But will they satisfy our human goals of exploration? This chapter addresses the question of whether our robotic spacecraft will really be extending the human presence in space.

Having spent almost a half century exploring planets—all done virtually (I have never left Earth)—I find no problem defining human exploration independently of the distance between the place being explored and the human. I remember soon after the *Viking* landing on Mars walking up to the director's office at JPL and seeing, for the first time, a large wall-sized photograph of Mars taken by *Viking*—the first image of Mars taken on the ground. It took my breath away, literally. I felt I was in the picture, seeing a world so obviously alien to ours and yet eerily like Earth. This, of course, was decades before computer animation, desktop image processing, creation of virtual worlds, etc. But even then the feeling of "being there" was possible because of the extraordinary imaging.

The evolution of robot capabilities is proceeding much faster than that of human capabilities. Imaging (photography) dramatically changed when it evolved from analog to digital, and each year now sees extraordinary advances due to electronics miniaturization. The *Viking* lander cameras were about 1.5 million pixels (approximately 3,000 × 500 pixels) and weighed about 7.3 kilograms. As I write this, a new cell phone has been introduced with 41 megalpixels (7,712 × 5,360) weighing 160 grams (including its smart computer).

Cameras provide one of the five human senses. To extend the human presence we need to consider the others. Hearing can be enabled by equipping the spacecraft with a microphone and then recording sounds. Just as human seeing is defined by the rather narrow range of visible light (roughly with wavelengths of 390–700 nanometers), human hearing is also defined by the narrow range of audible frequencies (approximately 20–20,000 hertz).[1] Equipping our robots properly, we can extend seeing and hearing to a broader range of wavelengths so that they can see in the ultraviolet and infrared, and hear at ultrasound and infrasound.

Equipping spacecraft with cameras and microphones seems like a pretty obvious idea. Not so. At the beginning of the space age scientists pooh-poohed carrying cameras on spacecraft as being only "public relations." They feared it would take the place of "real" science (physical measurements of particles and fields in the interplanetary environment). But exploration won out over science — i.e., the acceptance of the idea of looking for the unknown even without well-formulated hypotheses and experiments, giving the human brain a chance to interpret patterns instead of just analyzing data. Exploration and science do not fight each other but rather are mutually supportive — who can forget how images of Saturn's rings or Io's and Europa's surfaces changed scientific understanding about planetary dynamics. You would think that the camera experience would have made it easy for scientists to accept the idea of flying a microphone to hear putative Martian sounds on landing spacecraft. But in fact they resist it, on the same grounds they once resisted taking pictures, saying that we can predict what we will measure. That is why we want to try!

The Planetary Society privately funded a microphone on the ill-fated Mars Polar Lander in 1998 because we could piggyback it inside a Russian instrument, which was not subject to American scientists' peer review.[2] Science committees had opposed it. The Mars Polar Lander mission crashed during its attempted touchdown on the Martian surface. It wasn't because of the microphone. This mission was one of the first "faster, cheaper, better" projects directed by NASA administrator Dan Goldin, in an attempt to get more missions trying more things to fly to the planets, rather than putting all eggs into one expensive basket every 10 years or so. To his credit, he succeeded in increasing the flight rate (a lot), but he was forced to admit that perhaps he had pushed too hard on the "cheaper" part, forcing some cutbacks on the testing of the lander. Also to his credit, he made it possible to include Russian instruments on the American Mars spacecraft—something that did not work out for the instrument carrying the microphone (an atmospheric measuring lidar) but which worked out terrifically for the neutron detector, versions of which later flew on Moon and other Mars missions and contributed to important discoveries about the presence of water on these bodies. Since then, attempts to fly a Mars microphone continue to run into scientific resistance, and the idea still has to be "hidden" in the spacecraft electronics, seeking an opportunity to be turned on. Someday, I hope sooner rather than later, we will hear sounds on Mars—maybe only the wind rushing through the spacecraft appendages, maybe only the engineering noises of the spacecraft motors, or maybe a whole new physics related to electrical discharges and other atmospheric phenomena. If it is the latter, it may be up to the humans on Mars to listen.

A third human sense is touch. This is straightforward for a robot. We can easily measure properties of materials by touching them—sensing their roughness or smoothness, stiffness or elasticity, brittleness, surface density, etc. Touch data is easily gathered robotically. Taste and smell are much more complicated to instrument and measure.

But there is a huge industry that desires simple sensors to measure taste (liquids) and odors (gas)—e.g., food, beverage, and environmental monitoring, to mention three. Acoustic micro-sensors (measuring low-frequency waves) can distinguish physical properties of the liquids and gases. These data can be correlated to tastes and smells. Another approach is to use electrochemical techniques where a chemical reaction from the object induces an electrical response in a sensor. A robotic probe might have electronic tongues and noses, using one of these approaches in a miniaturized electronic circuit. Spacecraft have already measured chemical composition on planetary surfaces with instruments like gamma ray and neutron mass spectrometers, and other techniques with lasers are proposed for future missions. The chemical analysis provides us with digital data that can describe the tastes and smells of the sensed environment, just as the digital data from cameras describe images and digital data from microphones describe sound.

Thus we can robotically capture the five human senses, even to a much higher resolution than a human observer can. But, the human chauvinist notes, what about the sixth sense—robots can't measure or simulate human intuition. That may be true today, but progress toward that end is being made much faster than is progress on transporting humans to other worlds.

Artificial intelligence in computers is increasing dramatically—perhaps not as fast as the hyped predictions in the early years of the computer age but still very fast. Neural nets, "deep learning," and other programming techniques are moving closer to both brain simulations and adaptive learning by electronic circuits. Intuition is an ability to figure something out instinctively, essentially without thought. That means it is a hardwired process (or at least a firmware process) in us built on a combination of experience and genetic coding. Computers will build up similar experience as they make more and more computations covering all outcomes, and they are reaching the point where they will match the brain and our genetic code for the number of combinations that they can analyze. Programmers, in essence, train their neural network to recognize patterns, just the way the brain does in interpreting and even filling in the blanks with sensor data. The learned programs provide an experience base that in turn allows them to apply intuition (instinctive reaction to a pattern instead of analytical thought), that is, a sixth sense. I don't claim they will meet or beat humans in that, but they will do some, enough to help us in our exploration of other worlds.[3]

We see this happening in other fields, not just in space exploration. I already cited the military conduct of battles robotically, with drones now being used extensively for eyes, ears, and weapons. Drones are proliferating outside the military with many civil applications—some of which now create tension between goals of legitimate society, beneficial monitoring, and protection of privacy. Environmental cleanup and bomb and land mine removal are other examples of robotic devices enabling human participation to be distant and safe. In medicine, consider the *da Vinci*® surgery system, capable of performing coronary artery bypass, heart valve replacements, and hysterectomies through tiny portholes in the skin! It includes 3-D high-definition imaging, robotic arms and hands (with wrists capable of turning 540 degrees), automatic motion compensation, and an internal computer with input/output displays, all operated by a doctor sitting down at a workstation. The robotics, sensors, and input/output controls are also getting smaller. The *da Vinci*® system has little artificial intelligence—it is telerobotic and human controlled. But more and more computational power is included in the robotic and sensory parts to permit the human control to remain at the highest level of decision-making. We're years, maybe even a decade or two, away from this kind of telerobotic control being able to be done remotely (e.g., with the patient on Mars and the doctor on Earth), but developments in artificial intelligence allowing us to move in that direction are proceeding faster than are developments to transport doctors to other worlds. The same will happen for spacecraft—the exploring vehicle, even light-years away, will conduct intelligent exploration, analyze and synthesize results, and then over decades and centuries, beam back the information to the humans at home.

Another capability that characterizes, and hence extends, human presence is adapting: interacting with the data taken and getting new data, in essence automating the human observer's ability to pursue a question based on the answer to his or her earlier one. Some of this capability can be preprogrammed. We see this on websites now all the time: answer a question and depending on that answer get more questions in one direction or another. My rather simple e-mail "junk" filter quickly learnt a lot from my habits and now does a great job intuiting what to filter. But going further than that, we can think about in advance how we will use the aforementioned neural nets and other techniques of artificial intelligence to enable our onboard computer, as small as it is, to "learn." In addition to the near-instantaneous automated learning, our robotic spacecraft will have some ability to interact with us here at home and make us part of its learning. Even over interstellar distances, communication times are only a small part of the probe's lifetime. After 25 years of flight time, we can expect to be able to upload some very new programs and ideas into our spacecraft computer for it to make new measurements and experiments wherever it is flying. One whole new area that we can speculate about now is the use of 3-D printers—devices that can be programmed to make things. We might, by then, be programming them to make nano-things or biomolecular things, machines and new sensors that we can add onto our spacecraft payload. We'll have to use the raw materials we load up at the beginning of our mission—although some far-out thinkers will suggest extraction of space resources along the way to use on our automated craft. I think that might be possible in this millennium (i.e., in the 2001–3000 time period) but not in this century. My visionary horizon isn't as long as a millennium—I can't think that far. Two hundred years gives me a headache. But even in this time scale, I easily foresee 3-D printers programmed to adapt to onboard measurements to make new things, perhaps not in a totally closed feedback loop but in a pretty tight one.[4]

Even now in the infancy of first-generation 3-D printers, scientists at Princeton University have combined living tissue and a metallic material to create a bionic ear with a range many times larger than that of the human ear. The printed cells and nanoparticles to produce a metallic coil are imbedded in biological cartilage.

We have focused on the capabilities of the robot to sense data, make measurements, and be controlled, and we have seen that the capabilities for doing this at distant worlds is growing dramatically. But that is only half of the story of extending human presence. It's the half that has our extensions operating at distant worlds. But we aren't there, and therefore we can't really assert our presence as human. Only when the information comes home—where we can turn it into human knowledge—can we claim to have truly extended the human presence.

Processing the data taken by the spacecraft will be done by the onboard computers miniaturized from today's sizes through nanotechnology and biomolecular engineering. Some of the information will be used in the aforementioned feedback loops to reprogram the spacecraft to gather other data, and some will be coded into information for transmission home for human consumption. Nanotechnology will enable packing more brain into smaller volume. The properties of materials at the nanoscale (that is less than 0.1 micron, one-tenth of one-millionth of a meter, 1,000 times thinner than a human hair, and about one-fifth the wavelength of light) are very different than bulk materials—they behave and interact with particles and fields in novel ways. Scientists are learning to control the nano-materials and measure their interactions and thereby deduce information about the composition and structure of the environment in which they are placed. Packaging all of this into spectrometers, atomic microscopes, laser interferometers, and other instruments to create payloads for nano-spacecraft is a challenge. But seeing how the technology has evolved in the past 25 years gives one confidence that the challenge will be met. Let's take some snapshots of this evolution, say in 1980, 2000, and 2020, as shown in table 3. The year 2020 isn't very far from now—thinking about the end of this century would be truly way out.

More generally, we can track the evolution of the capability of simulating the human brain with computers. Artificial intelligence began as an idea in the 1970s with the programming of expert systems (essentially finding things in databases). After ups and downs in the 1980s, alternately predicting fantastic breakthroughs and then debunking hype, IBM developed a computer and program that beat Garry Kasparov, the greatest chess player in the world. Still, the computer was essentially equivalent to a few thousand neurons compared to the brain's hundred billion (or so). By the mid-2000s, the Defense Advanced Research Projects Agency's (DARPA) grand challenges for robots to drive autonomously through traffic and on highways were met, and in 2010, another IBM computer defeated *Jeopardy* game show champions. Still, the brain complexity of the computer was below that of a rat (about 100 million neurons). By 2013, we achieved the rat level, and we probably will reach the human brain level in another

Table 3. Snapshots of technology evolution, 1980, 2000, and 2020

Technology level	1980	2000	2020
Memory size (the smallest feature of random access memory)—microns	1	0.1	0.02*
Microprocessor clock speed—MHz	2	500	20,000
Processor performance—millions of instructions per second (MIPS)	0.05	1,500	>100,000
Power required for computation—watts/MIPS	2	0.0005	~10-6
Memory capacity—kilobits/$	10	10,000	billions
Supercomputer performance—floating point operations per second (FLOPS)	–	1012	10^{18} (~0.1 human brain)
Power required for one MIPS—watts	1	0.0005	<0.00001
Genetic data sequence/year (sequences)	<1000	5 million	>100 million

*This is approximately one billion gigabytes per cubic centimeter!

decade. Packing it into a spacecraft may take another decade or two after that: about the same time as we are likely to finally put people on Mars (and very long before we are capable of transporting them farther).

Spacecraft are already evolving into smaller and smaller packages in ways very different than we can imagine by looking at our current fleet of interplanetary spacecraft. As of 2013 there were at least a half-dozen startup efforts seeking to build "personal" spacecraft, or spacecraft-on-a-chip, of which dozens or even hundreds can be packed into CubeSats (see chapters 4 and 5) and operated from smartphones or other mobile computing devices. I don't suggest that these spacecraft will themselves be capable of navigating to deep space, but I do think their technology presages what we might use in the autonomous nano-spacecraft of the future, which will be an interstellar craft.

Because of their size, the spacecraft will operate at very low power levels (see table 3), and thus even over interstellar distances, we can expect to receive bits of information, which over days, months, and years become megabits. In turn, these data will be processed on our late twenty-first-century computers with artificial intelligence and neural nets (or their derivatives) to create virtual worlds, quite likely in three dimensions (we already have 3-D TVs), mimicking the real worlds in our computers. Our probes, interacting with the environments of other worlds, will be extensions of our human presence, and the data they return will allow human exploration of them in their virtual re-creation. This has begun already with data from Mars taken by the *Curiosity* rover being loaded into an Oculus Rift (a virtual reality device) and then explored by a person on an Omni multidirectional treadmill.[5] So far only imaging data is in the Rift (permitting the person to walk on the Martian landscape), but as soon as the JPL Mars rover scientists tried it, they began thinking about how other data (e.g., chemicals, temperature, wind) might be included.

Another far-out suggestion is to communicate, via radio or optical means at the speed of light, genetic codes from Earthly organisms to a distant spacecraft or vice versa to send a discovered "foreign" genetic code from a distant world to Earth. The idea of transmitting life digitally sounds a little like teleportation—although the idea, proposed by Dr. Craig Venter in his 2013 book, *Life at the Speed of Light: From the Double Helix to the Dawn of Digital Life,*[6] is more to copy the genetic code into a computer in order to transmit the information. Ideally, we could take that copied code and sequence it to make transgenic organisms and to synthesize the foreign molecular structures—whatever they are. Ideally, to be sure; we really don't know yet if that will be possible. We haven't synthesized life on Earth from DNA nor done any DNA transfer without a host organism. It does seem, however, that biologists (particularly Venter) are on that path and that such technology will be in fact on the same (or even faster) time scale as sampling biologic molecules on other worlds. In the initial workshop of the 100 Year Starship™ (chapter 5), Venter urged us to think about molecules in our spacecraft payload programmed to interact with an alien planet's atmosphere and then transmitting the results back to Earth via radio or optical communications.[7] This is clearly extending human presence without the human present. While I am sure we are a long way off from reducing life to pure numbers, I am pretty sure that harnessing digital biology (and biomolecular engineering) is a faster process than shipping human bodies on interstellar travel. A group of NASA scientists, notably led by Dr. Chris McKay (a former member of the board of directors of The Planetary Society) is testing Dr. Venter's idea out on Earth as a means of someday sending biologic information from Mars to Earth. As argued in the next chapter, I still believe we will send humans to and from Mars and possibly even settle humans there, but this kind of evolving technology in both electronics and biomolecular engineering will negate the need and the urge to send humans any farther.

Our current younger generation has already started the transition, making the tele-presence/telerobotic/virtual exploration experience just as much human as the old-fashioned "being there." Succeeding generations will take it even further, and thus, interstellar flight will be accomplished by robotic space explorers not just first but always. Some find that negative, but that may be a failure of imagination, no more negative than my grandson telling me how social and involved with friends he is, even though he hasn't actually been with anyone outside of his house for days. He is socially networked—and someday, he or his children will be networked to other worlds the same way.

Ray Kurzweil has taken these ideas to the extreme in a *singularity* theory,[8] but I fall short of that. Kurzweil emphasizes that he believes humans will evolve to some mixed state of biology and electronics and will send their intelligence to explore, whereas I, much more mundane, emphasize only our still-distinct robots sending back data to us still-distinct humans to manipulate in virtual exploration and experience. He seems to merge digital biology and living electronics into one. Maybe he is right, or maybe we are only headed for a *duality*, instead of a singularity—of robots and humans.

I find the accusation of being negative about the future really annoying.[9] Negative is in the eye of the beholder. To assert that human space travel is limited does not mean that either humankind or humanity is limited. Spaceflight is but one means of extending the human presence and of human evolution. And it is a cumbersome and inelegant one, compared to what is going on with other means of extension and evolution. To envision human interaction with robotic nano-spacecraft profusely disbursing through the galaxy investigating and perhaps interacting with hundreds (if not thousands) of other worlds and possibly other life is, to me, pretty uplifting and quite positive. Add to that the human settlement and evolution on nearby Mars, and it becomes even a more optimistic future. To prove this, we next consider the possibilities for humans on Mars.

CHAPTER EIGHT

Creating a Multi-planet Species

Humans to Mars

We all dwell in a house of one room . . . and are sailing the
celestial spaces without leaving any track.
—JOHN MUIR

In the previous chapters we have described how robotic probes will enable us to create a multi-planet species. We will be virtually present on and interacting with other worlds. Humans will not have to adapt or survive on the extremely extreme environments of the outer solar system—our probes will fill up virtual worlds with real data to explore. Nano-spacecraft will even be able to carry out interstellar precursors, perhaps even in this century (as described in chapter 5). Already, interplanetary CubeSat missions are being developed, and solar sailing is reaching a new level of readiness, with several missions planned in the next few years. Perhaps we will even reach the solar gravitational lens focus on a mission dedicated to choosing humankind's first interstellar destination. In addition, a new generation of robotic spacecraft will be landing on moons of Jupiter and Saturn, specifically on moons of astrobiological interest: Europa, Titan, and Enceladus. With the technology advances of electronics, information processing, and biomolecular engineering that we described in chapters 6 and 7, we can envision three-dimensional holograms (or their informational equivalent) of these other worlds available on Earth from the spacecraft data—not only to observe and immerse ourselves into but also to interact with. These data in those holograms will allow us to virtually cut a hole in the ice of Europa, moving underneath and sampling its constituents, just as if we were physically there.

In the previous chapter we mentioned Oculus Rift being used for Mars exploration, suggesting we are almost there with such virtual exploration.

Our pace of doing this might be slow. At The Planetary Society in the early 2000s, we were lobbying NASA and Congress to approve a Europa mission to launch by 2015. Even with National Academy of Sciences endorsement, international agreements with Europe for cooperation, and congressional approval (albeit without sufficient funding), the schedule for an orbiter mission with no lander has been slipping every year. A scaled-back multi-flyby by a Jupiter orbiter is now being pushed for launch in the early 2020s. When we can send a lander there to probe beneath the ice is anyone's guess—mine is 2050. Still, that will be decades, if not centuries, before human space travel could reach the outer solar system, let alone be sustained. The first humans will likely get to Mars before a robotic probe lands on Europa, and certainly before one lands on Titan.[1]

Learning and adapting to this data from other worlds will change our perspective of our own world. It will also provide real information that helps us deal with our environment, manage our resources, and create new areas of human endeavor. Even if we haven't left Earth ourselves, we will be a solar system species, much in the same way that globalization of industry, agriculture, and trade has made us a global species even for those who have not left their home country.

That said, it is not clear (either psychologically or socially) whether we will really regard ourselves as *on* other worlds or whether we can count ourselves as a multi-planet species if we are only physically on one planet. Extrapolating today's technological and cultural trends leads, as I argued in the introduction, to finding our virtual presence to be more "real" than extrapolating our ideas about physical travel to other worlds. However, even if that is correct, there is still one component lacking from that creation of a multi-planet species, that is, having resiliency against catastrophic species extinction here on Earth. If we are not physically able to live on another world, then all our eggs for survival of the species are in one basket. Earth has suffered and will suffer many species extinctions, but humans are the first species to be able to think about and deal with it. Asteroid impact, climate changes, war, disease, and combinations of these are among the factors that threaten our species. Some may be under our control and some may not. I am (sadly) amused at times by the climate change deniers who want us to ignore reality—because they are worried about who will be assigned blame for it. It doesn't matter what causes climate changes—humans, sunspots, passing galaxies, or even worms—the fact is that unlike in the past we have hundreds of millions of people who will die if we ignore it. Of course not ignoring it means we might have to take environmental actions that offend some ideology-based thinking. Interestingly, in military defense matters some of these same people are always arguing we must plan for the worst case and be fully prepared (and, in fact, the U.S. Department of Defense has now categorized climate change as a national security threat).

No matter what occurs—climate change, nuclear or biological war, pandemic, or asteroid impact—all could lead to global catastrophe and threaten human survival. Having another world is prudent, but would it make us complacent? Political movements can be strange, but it is hard to imagine the one that says, "Ease up on saving Earth, we always have Mars." Space solutions to Earthly problems will always be more expensive, more difficult, and less certain than Earthly solutions. But still, there might be cases where Earthly solutions are just not enough—the large asteroid impact is the most poignant example.

Prudence, and human survival, drives us to create another home for humankind. Between here and the stars there is only one possible home: Mars. It alone has accessible oxygen and water and might support us humans. And, it alone can be reached before we evolve so far as to completely accept the virtual worlds created here on Earth, in lieu of going to other worlds.[2]

Don't let me pretend Mars is clement and nearby. It is not. It is a frigid, hostile, bone-dry, likely lifeless planet that will take a lot of adapting to for any human activity. It may warm up to near 0°C (freezing temperature) at the equator in the summer—but on average the temperatures are –60°C. Mars's atmosphere is poisonous: 95 percent carbon dioxide (compared to 0.04 percent on Earth). We do see evidence of copious water having run on Mars, and we have now even observed places where Mars water has bubbled to the surface in recent times, but let's not overstate it. Mars's surface is drier than the Earth's driest deserts, and if the water that we hope to find beneath the surface (although we do not know how far down) were extracted to the surface, it would instantly sublime—unless we immediately captured it in a pressurized container. If cold, poisonous, and dry aren't enough to dissuade you from living there, then being fried by solar and cosmic ray radiation should be; Mars lacks a thick atmosphere to shield us from high levels of radiation.

With all this, I still think Mars is a future home for humanity. For all its hostility, it is the only other place we can both reach and adapt for living. Culturally, the leap to Mars is even easier. As executive director of The Planetary Society, I frequently gave public lectures about planetary exploration. I would ask the audience, "How many of you think humans will go to Mars?" and most hands would go up. Then I would ask, "How many of you think that humans have *already gone* to Mars?" and still a lot of hands went up. (And when I asked how many thought Martians have come here to Earth, that too would cause hands to raise!) Many in the public believe that Mars and Earth have already exchanged life.

In the late 1980s, I was working with an international team trying to develop a balloon that could fly in the thin Martian atmosphere. A Mars balloon is a neat idea. With it (on a robotic mission), you could gain very high-resolution imagery over a huge swath of Martian territory—much more than the few kilometers traversed by our wheeled rovers. To test the feasibility and possible instrumentation for a Mars balloon, we conducted hot-air balloon flights with simulated Martian measurements and instruments. That we were doing this in the 1980s for a Soviet Mars mission with American, French, and Soviet scientists was a political challenge—and a political adventure. We conducted tests in the Mojave Desert in California and then decided to try it in the Soviet Union (still a nation in 1988). For some reason (having to do with enthusiastic personnel), we ended up flying hot-air balloons in Lithuania on a Soviet military airfield. Together with my colleagues Tom Heinsheimer (an aerospace engineer and serious amateur balloonist mentioned earlier in this book), Slava Linkin, and Viktor Kerzhanovich (planetary scientists at the Space Research Institute of the Russian Academy of Sciences),[3] we flew out of the airfield designated area, to the chagrin of the officials and other team members on the ground, and over a river to another part of the country—more than a two-hour drive from the nearest bridge. We landed next to a cow in a rather rural area. A group of local people (from neighboring farms) gathered around us. I ran up to greet them and explain ourselves in what I thought might be pretty good spoken Russian. I said that my colleagues and I were an international team working on space missions to explore Mars and that we were flying in the balloon to test ideas about making measurements on Mars. All went well for a couple of hours (we were offered food and drink in a small house from a very nice babushka whose farm it was). But after they had a real Russian-language conversation with Slava Linkin, I learned that the local folks thought I had explained that we had flown by balloon from Mars and happened to land there! What was remarkable to me was not that the story was nuts but that they accepted the idea that Martians had landed—not for the first time on Earth. The public does not have too much trouble accepting Earth-Mars cohabitation.

Mars beckons: by my count, 55 spacecraft as of 2014[4] have been launched from Earth in the 55 years of the space age from five nations (the United States, the Soviet Union/Russia, Japan, China, and India) and the European Space Agency (ESA). Of these (again by my count), 37 reached

Figure 14. Mars beckons: landscape from Mars *Pathfinder*. NASA/JPL.

Mars and were successful. The United States has landed successfully seven times, and the Soviet Union had two successful landings—but only with very limited mission success. Despite many failures and setbacks, the lure of Mars continues. The number of craft reaching Mars is comparable to the number of ships that reached the Americas during the whole sixteenth century. Of course the latter were manned and included some extensive land expeditions and wars of conquest, whereas the former were passive observers—extensive in their own right. The surface area of Mars (remotely observed now by the orbiters sent there and in a few spots by the landers) is equal to that of the land area on Earth.

It is worth thinking about why there is such a lure for continuing Mars exploration. It isn't the taking of slaves or the promise of gold like what drove sixteenth-century American exploration. Nor is it that political leaders made some strategic decision to race to or conquer Mars. They won't even commit to explore continuously—each decision is considered a "one-off." Yet through the ups and downs of budgets, and even through the collapse of communism and the end of the Soviet Union, Mars exploration is renewed, and the goal of even human exploration pulls on the program. There is little effect on daily life, economics, or even fundamental science or technology from Mars exploration—why then does it stay as a continuous effort? Why do nations, even with budget difficulties and developing and struggling with larger socioeconomic problems like India and China are, start developing Mars missions almost as soon as they becoming spacefaring? In part it is because Mars represents a symbol of national capability and leadership. But, I submit the answer is deeper than that (many things are national status symbols) and that it lies in the public interest, often latent, in answering the big questions of life: who are we, where do we come from, where are we going, are we unique or part of a cosmic population? I find it not surprising that countries and populations are willing to spend a tiny percentage of their budget[5] on such questions represented by the exploration of Mars because society has spent many times that percentage over history on those same questions through folklore, stories and religion. Now we are doing it with science and technology. Sending humans to Mars is a manifestation of human unwillingness to live with limits.

Mars is the only place that we can reach and search for answers to those big questions. And Mars never disappoints—just when we think we know something, it changes. The early Mariner missions seemed to indicate it was a dead planet, later ones indicated it was an active place with recent running water, and then *Viking* found a toxic, sterile surface inhospitable to life. But, following the water, later missions found many possibilities suggesting life could have existed in the past and might, even with that sterile surface, still exist in underground niches. Whatever the possibility is of past life, environmental understanding of Mars suggests that it could be a habitable world and even a possible future habitat for humanity. Of course humans on Mars would need a lot of environmental protection—from radiation, from a thin and toxic atmosphere, from frigid temperatures—but we already do a lot of environmental protection and adaptation here on Earth,[6] and we are going to have to do more. Doing this on Mars, which has the elements of life already there, is certainly possible. We can imagine living on Mars bases and habitats underground on Mars, protected from radiation, near accessible water, and maintaining a livable environment.

How to do this has been brilliantly described by Kim Stanley Robinson in his science fiction trilogy: *Red Mars*, *Green Mars*, and *Blue Mars*.[7] The colors refer to now (Red), an environmental and adaptation transition (Green), and finally an Earth-like settlement (Blue). The process of changing the planet from its current inhospitable state to one that could support humankind in a natural way is called terraforming. It consists of climate change through engineering—sort of the reverse of what commands so much attention on Earth. Here we are worried about the deleterious effects of climate change on life: things like increasing carbon dioxide and other greenhouse gases resulting in a warmer planet with melting ice caps and glaciers. On Mars our descendents might want to deliberately warm the planet to create a more supportive atmosphere, as opposed to our inadvertently warming it here on Earth due to carelessness and environmental mismanagement, creating a less supportive atmosphere. Living in subterranean habitats might be possible in a century or two; terraforming is at least a millennium-long process.

Figure 15. Mars base with solar sails. Ken Hodges, NASA/JPL.

Figure 16. Terraformed Mars. Michael Carroll.

Robot exploration is limited. Of the four rovers sent to Mars, *Sojourner* traveled a total of 100 meters, never more than 12 meters from its landing spot. *Spirit* traveled 8 kilometers before one of its wheels froze and became stuck. *Opportunity* is still operating; it has traveled 40 kilometers (in approximately 10 years). *Curiosity* also is still operating, having traveled about 700 meters in the year after its landing, and is now headed to its next destination, Mt. Sharp, an approximate 8-kilometer traverse. These are impressive accomplishments all. But they also emphasize how little of Mars has actually been explored and how slow the pace of our robotic rovers is. *Curiosity*'s (the fastest of these rovers) top speed is approximately 140 meters per hour. A typical human geologist on Mars might walk 3,000–5,000 meters per hour (20–35 times faster), even while making observations and notes while moving. He or she will drive a rover even faster.

We are learning a lot about Mars, and we are indeed fortunate that Mars is being so regularly visited and explored by many nations. But even if budgets permit our most ambitious plans (and it is likely they won't), it will be many decades before we have any significant amount of Mars explored and characterized. I am an avid proponent of vigorous Mars exploration—by all spacefaring nations,[8] by robots and humans, from orbit and on the surface. As executive director of The Planetary Society, I considered myself a paid optimist advocating space exploration. As vigorous as exploration can be, and as optimistic as I want to be, I cannot imagine Martian settlement even being considered for another century. A human Mars base should be achievable in this century if our dreams are realized and there are no major setbacks. In the following I will describe those dreams and how I hope they can be realized for both robotic and human spaceflight. But the important thing to realize is that even in our most optimistic scenarios human "colonization" of Mars will take centuries—*if it occurs at all*! And during those centuries, robotic technologies as described in the previous chapters will be advancing faster and farther and deeper into space. By the time humans settle on Mars, more distant human travel will be conducted without humans onboard.

I emphasized "if it occurs at all." That is a big IF! As stated, I am optimistic; I believe progress and exploration will continue unabated, or at least only briefly abated. It may not. It is easy to imagine many interruptions: war, economic crises, disease outbreak, social unrest, environmental crises, and political disinterest, to name some. I also believe we are in a space race, not the kind that spurred two competing nations to goad each other to move faster but one that affects human attitudes: a race between robots and humans. This is not the trite one where advocates for or against human spaceflight argue about which is better or more important. I mean one in which human capabilities and social attitudes stay competitive with robots. If we choose (politically or culturally) to reject human space exploration, then it seems likely that our attention will wander to the point where we are satisfied with robotic exploration—not unlike the way a younger generation today is satisfied with text messages rather than talking or (dare I say it) interacting by touching.

It is because the pace of robotic and virtual exploration will enable human interaction on distant worlds without humans being there that I conclude that humans will never go beyond Mars. At the opposite end of the scale, I also said earlier that humans will not be satisfied staying within the limits of Earth—our perception of species survival compels us to become multi-planet. That is the ultimate justification for humans to go to Mars. The danger of not sending humans to Mars is that we will become complacent with the evolving robot and virtual technologies. If that complacency overcomes the multi-planet species compulsion, then we are doomed. I do not believe that will happen—the question of human settlement on Mars should occupy for us for centuries, indeed even the rest of this millennium. After that, with our human presence fully extended, the robotic and virtual world victory won't bother us.

There is (if you will excuse the idiom) a fly in the ointment. To succeed as a multi-planet species, Martian settlement must be the goal. But it is not obvious that settlement would be achieved. We have not settled in the far more accessible, hospitable, resource-rich environments under the oceans or in Antarctica. I recall reading *Popular Science* articles and seeing cover pictures about cities being built under the oceans with people living off the food accessible there (fish and vegetation), extracting oxygen and other nutrients from seawater, and engaging in reasonable trade and commerce to the air and land above them. Not only has that not come to pass but also there is no direction or movement suggesting that it ever will. However, it has only been a short time (less than a century) since I read those *Popular Science* articles—too short to make a conclusion yet about the inevitability of colonization.

Searching for life drives us into the universe—as noted earlier it is really a search for ourselves. The drive toward Mars is additionally motivated by the visions of humans settling, or at least exploring, there. "When are we sending humans to Mars?" is still one of the main questions we get when we present popular lectures and hold discussions with public audiences. It also dominates space conferences, policy studies, and periodic reconsiderations of space programs. Ever since the beginning of the space age, humans to

Table 4. Current Mars missions

Spacecraft	Type	Nation/agency	Launch	Arrival
Odyssey	Orbiter	United States	April 2001	October 2001
Mars Express	Orbiter	ESA	June 2003	December 2003
*Opportunity**	Rover	United States	July 2003	January 2004
Mars Reconnaissance Orbiter (MRO)	Orbiter	United States	August 2005	March 2006
Curiosity	Rover	United States	November 2011	August 2012
Mangalyaan (MOM)	Orbiter	India	November 2013	September 2014
Mars Atmosphere and Volatile EvolutioN (MAVEN)	Orbiter	United States	November 2013	September 2014

*A companion, *Spirit*, was launched in June 2004, but ceased working in March 2010.

Mars has been a goal set about 30–40 years in the future. That number seems pretty constant. In 1965, NASA technical memoranda and studies envisioned human Mars missions around the year 2000. In the 1990s, we thought it might be around 2020–2030. Today, various roadmap charts (and President Obama's speech in April 2010) envisioned it in the 2040s. It is easy to take a cynical view and say it will remain 30–40 years in the future forever, or at least for the rest of this century. That would ignore the truly steady progress we are making in exploring Mars. Orbiters, landers, and rovers are now making their way there, not just from the United States but also from other spacefaring nations. Table 4 lists missions currently active at the red planet, and table 5 lists those under current construction.[9]

And that's not all. Russia has announced it will make another attempt to fly its Phobos[10] sample return mission in 2022, again with an orbiter from China, and Japan is studying an orbiter/lander mission also for 2022. NASA has begun studying the development of Mars missions throughout the 2020s, including sample return ones.

It is important to understand that all of the planned Mars missions are scaled-down compromises of earlier plans by the space agencies for more ambitious exploration. The original plan was for the United States to have a major role in the ESA's ExoMars missions and its own rover in 2018. Russia, NASA, and ESA (at least) should be working more cooperatively

Table 5. Planned Mars missions

Spacecraft	Type	Nation/agency	Launch	Arrival
ExoMars 2016	Orbiter + test lander	ESA with Russia	January 2016	September 2016
InSight	Lander	United States	March 2016	September 2016
ExoMars 2018	Lander + rover	ESA/Russia	TBD	TBD
Mars 2020	Rover	United States	TBD	TBD

and coordinating development for the most desired science goal on Mars—returning a surface sample for detailed chemical and possible biological analysis in Earth laboratories. This has been talked about for years—in fact, I even led a Mars sample return study when I was leader of the Mars Program at JPL in 1977—and that was not the first NASA Mars sample return consideration. Prior to the failures of the Mars *Climate Orbiter* and polar lander missions in 1999, NASA and European countries were working on a Mars sample return mission plan for 2003–2007. The aborted 2016/2017 joint NASA-ESA plan was to collect samples for a 2020 or 2022 mission to pick them up and return them. The chief technical advantage of these multiyear cooperative plans was to spread out the cost and risk over many years and many nations to make it fit even in constrained budgets. But the budget bureaucrats don't like multiyear plans because they don't like multiyear obligations. That of course leads to piecemeal plans—less than we hope, but still continuing exploration. The planned missions listed in table 5 still convey optimism while countries bang back and forth in economic crises.

Mars sample return is sometimes called the "holy grail" of robotic space exploration—always being sought, never quite realized. It's an exciting mission with multiple spacecraft to be coordinated for launch, cruise, orbiting, landing, roving, sampling, and returning. Staging is a principle of efficient space vehicle design. The most inefficient way to launch a payload from the surface of Earth is with a single-stage rocket trying to carry all the necessary propulsion mass to orbit. Better to break up the propulsion into stages so that you can drop off the mass of each stage when the propellant is expended. Otherwise you have to carry the dead weight of the used propellant tanks with you as you go to higher distances. That's why launch vehicles have multiple stages. There are still some proponents of single-stage to orbit who are attracted by the theoretical simplicity and potential economy of a one-stage vehicle—but no design has yet proved practical for implementation. The same multiple-stage principle applies to interplanetary missions with multiple phases as with Mars sample return missions that requires Earth-Mars cruise, Mars orbit, Mars entry and landing, Mars liftoff, Mars orbit, Mars-Earth cruise, and Earth entry. In theory you could try to do that with one vehicle, but it would be not only very inefficient but also extremely sensitive to design calculations. A small change in one part of the full system would cause major mass penalties and changes throughout (see appendix). Thus, good design leads to different spacecraft stages—multiple vehicles for the mission. This in turn leads to natural international cooperation (with nations each being able to contribute a part)[11] as well as to separation of the mission costs into smaller parts that can be done over several years. For example, in one year the surface sampling vehicle could be launched. After it gets to Mars and is operating, succeeding missions could launch the rocket to go to the surface, retrieve the sample, and bring it to the Mars orbit, where it might be transferred to still another vehicle to bring it back to Earth. Splitting the cost among several nations and over several years might make the multibillion-dollar mission affordable for each participant, and thus affordable as a whole.

Cost has been one big impediment—every time a science or engineering team gets a Mars sample return proposal completed, they hit the budget wall. In addition to all those vehicles (and their requirements), the mission has tacked on to it heavy launch costs (for all the parts) and a major cost for sample recovery and handling. The cost of safe and reliable sample recovery, containment, retrieval, storage, and environmental protection is large because extreme care is required to deal with the danger of contamination—on the one hand contamination of the sample's purity and integrity and on the other hand possible (however unlikely) contamination of the Earth from the alien sample. All Mars sample return proposals failed because of an inability to deal with the costs.

The chief arguments in favor of bringing back a sample from Mars (over more in situ investigations) are the superior capabilities that can be brought to bear on the analyses in Earth-based laboratories unconstrained by mass and power spacecraft limits and the particularly unique age-dating that cannot be precisely done on board spacecraft. A sample brought back to Earth can be analyzed with more modern and capable instruments than those that would be packaged in a spacecraft. On the Mars Science Laboratory mission (with the *Curiosity* rover), some clever age-dating was done by combining results from several geologic and geochemical experiments, but age-dating under Earth-based microscopes would be far more precise.

As much as I personally would value robotic sample return, I am mindful of our lunar experience. The Apollo astronauts were far more capable sample return enablers than the Russian robotic Lunokhods. If we are doing a human-crewed mission soon, even if it is a decade or so later than the robotic mission, then we might argue that a robotic Mars sample return is a needless expense since the human mission would do much more, both with more sophisticated sample selection and heavier payload capacity for returned samples. Scientists want to do precursor robotic sample returns, but experience to date suggests that the science program alone can't afford that luxury. Engineers also want to do a precursor robotic Mars sample return for safety and reliability prior to risking humans on a Mars mission. That is probably a good idea and might be practical, if not for the aforementioned cost of science requirements and sample handling. A precursor round-trip mission could be carried out at much lower cost if the science of sample collection and handling were deferred to the human-crewed mission.

Earlier I mentioned a human-robot space race, mostly for the hearts and minds of the public. I would gladly forego the robotic Mars sample return if it increases the human drive to go to Mars. Multiple landers and rovers sent to diverse areas of Mars for in situ exploration may be a far better and more practical precursor for human exploration. Whatever we learn about Mars from our robotic missions will only whet our appetite for getting humans there. My expectation is that the missions currently being worked on will continue to set a stage for multilateral, international collaboration for a series of Mars landers and rovers to different sites in the 2020s and early 2030s and that these will enable selection of a prime landing site candidate for humans on Mars by 2040—perfectly coinciding with the long-delayed human flight capability for missions beyond the Moon (see chapter 9). The human landing on Mars would most likely be preceded by humans flying around Mars without landing, perhaps visiting the Martian moon Phobos, building up their long-duration interplanetary flight capability. Hopefully this can happen in the mid- or late 2030s.

A scenario like that would allow humans to win the race to become a multi-planet species by or at least during the twenty-second century. From Mars and Earth we can then extend our reach beyond human body limits into the solar system and beyond. We will, by then, have the capacity and technologies for human extension without the human presence.

That is the scenario I like. But there is a significant amount of opinion that until we prove there is or was no life on Mars, humans should be super-cautious and not contaminate the planet. People with that opinion say we should only send sterile robots there for the time being. That leads, however, to the following conundrum:

1. You cannot prove a negative in science. Specifically, you can never prove there is not life on Mars. (At best, you can prove it is unlikely.)
2. Thus, there will always be uncertainty and a body of opinion that says there may be life on Mars, so leave it alone.
3. Therefore, never send a human to Mars.

I do not subscribe to this argument. I believe the gain of exploration and new world settlement will outweigh the risk.

Of course, even clean robotic spacecraft are not perfectly clean, so Mars might have already been contaminated. Its surface, however, is highly toxic, and it is likely that any contamination introduced there is long dead. Furthermore, the constant bombardment of cosmic ray and solar UV radiation makes the surface sterile.

There is another race besides that of humans and robots. There are government and private interests competing to advance human exploration. There are now several proposals (at least in the Internet chatter stage) for privately funded missions to Mars. Elon Musk, the founder and leader of SpaceX who has already built successful Falcon rockets and commercially serviced the International Space Station, says that his *Dragon* capsule intended to carry humans to and from Earth orbit is capable of soft-landing a payload (without a human crew) on Mars. It might even be able to carry sufficient payload for a sample return mission. A group of NASA scientists are studying the idea of a "Red Dragon" mission that could be launched by the heavy-lift version of the Falcon even in this decade. Would Musk do that privately? That has been suggested, but there is no evidence of it yet, despite his avowed enthusiasm for Mars exploration. Even if he did, that is a long way from a privately funded human-crewed mission to Mars.

Two groups gained some small (and skeptical) press attention recently by claiming they would do privately funded human missions. One, led by space tourist Dennis Tito, is for a round-trip flyby mission of Mars (with no landing), carrying a crew of two. It's a neat idea, taking advantage of a particular planetary alignment (in the year 2018) enabling a low-energy 500-day round-trip mission. He has formed the Inspiration Mars Foundation to raise the money to carry it out—at an estimated cost of (they say) about $1 to $2 billion. The second is called Mars One (based in the Netherlands). They claim they will start a Mars settlement with a one-way human-crewed mission in 2023(!) after emplacing a habitat on the surface there prior to the crew's arrival.

These private claims are audacious and not really credible. But they are statements of intent, and eventually (if not in the next decade or two), where there is a will there will be a way. It's all good—so long as they do not become considered as political alternatives to the far more likely and credible government efforts, worldwide, in Mars exploration. As noted in earlier chapters, precisely this sort of thing happened with lunar exploration after Apollo and with the Comet Halley opportunity in the United States in the 1980s. In both cases private groups asserted they would do missions if only the government would get out of the way. The government got out of the way and nothing happened.[12] I personally believe that government is the proven best way to organize great projects of exploration with broad society benefit. Waiting for private investors to do it with no return (or very long delayed return) on investment is an excuse for inaction.

Mars is not the only place of astrobiological (or putative life) interest in our solar system, and in fact, some astrobiologists assert it is not even the place of prime interest. They point to the water-rich environments on outer planet satellites, notably Europa around Jupiter and Enceladus around Saturn. That these frigid worlds have liquid water wasn't suspected until it was observed by spacecraft—and then scientists figured out that the interior of these moons were heated by tidal friction caused by the gravity of the large planets they orbit. If there is energy creating heat and water and organic material (abundant in the outer solar system), then life may indeed have formed out there. It may be more probable than on the desiccated surface and subsurface of Mars. But even with the possibility of indigenous life, these places are not places for humans to go. They are totally inaccessible and hostile—at least for time scales of hundreds of years. And, as argued repeatedly in this book and specifically demonstrated in the previous chapter, over that time scale human exploration will be done without humans being there. Our robots with the putative biomolecular and nanotechnology payloads and their super artificial intelligence will create those worlds for us to explore virtually here on Earth or perhaps in the comfort of our new Martian settlement. This will also of course be the case for the to-be-discovered astrobiologically interesting worlds around other stars.

There is a lot happening at Mars. The robotic program now includes several nations, and the number of landing sites continues to increase. Many more are planned in the next decade (i.e., from 2015 to 2025). The nations conducting these missions (the United States, Russia, Japan,

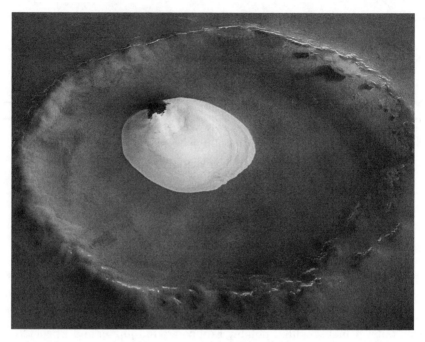

Figure 17. Mars of astrobiology interest; water ice in bottom of crater. ©ESA/ DLR/FU Berlin (G. Neukum).

Europe, China, and India) are also conducting and developing human spaceflight.

If even some of this goes well[13] we should see robotic experience at Mars and human spaceflight coming together early in the 2030s, and the human Mars goal will beckon even more strongly, leading to its realization by 2040 (the Bush and Obama administrations' goals), or at least by 2050. That is logical technically and scientifically, and I believe even economically (affordable within reasonable projections of today's space program budgets). But there is a dichotomy to be reckoned with.

The dichotomy is inspiration vs. politics. Mars looms in culture and public attitude as the human space exploration goal—driven by the questions of extraterrestrial life and the future of human settlement. But politically it is too big a commitment for the accountants in charge of budgets, and it doesn't deal with any immediate crisis, thereby keeping it in the future. We see this every time it comes up for political consideration. First it was the Wernher von Braun dream that inspired the Apollo program. However, President Kennedy's goal was more geopolitical and immediate—beat the

Figure 18. Europa, cracked ice surface showing water evidence. NASA/JPL.

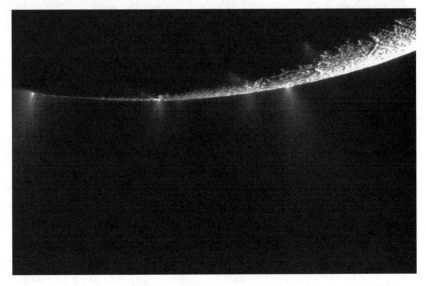

Figure 19. Enceladus water plumes. NASA/JPL.

Figure 20. Titan. NASA/JPL/University of Arizona.

Soviets to the Moon. As soon as that success was achieved, the political push for space travel diminished—fast. Then a high-level commission under Vice-President Spiro Agnew recommended it to President Nixon. That went nowhere and in fact led to a cutback in NASA that confined humans to low Earth orbit for the next five decades. A brief flurry of interest by President Reagan and Soviet Prime Minister Gorbachev for a human Mars cooperative project disappeared quickly, also in the face of much larger global events such as the dissolution of the Soviet Union. Presidential candidate Al Gore endorsed the joint human Mars mission idea but abandoned it quickly under President Clinton, where the focus was again on a near-term strategic goal: engage the Russian military-industrial complex in the International Space Station. The word "Mars" was congressionally banned from the NASA human spaceflight program for awhile in the 1990s. Both of the Bush presidents (George H. W. and George W.) proposed new programs for NASA with the goal of sending humans to Mars. Neither went anywhere. Indeed, both were resisted by NASA itself: the first time by pricing the Mars goal too large to be realistic and the second time by giving up on it entirely in favor of seeking to create a Moon base.

Never has the long-range view been enough to drive the long-range program. As I write this, we are in the midst of another presidential initiative — this one from President Obama who proposed skipping past the lunar base and creating a program to send astronauts into the solar system, reaching Mars in the 2030s. This time NASA supported the initiative, but Congress refocused it on building a new rocket with new government contracts — taking care of constituents instead of the space program. The funding for NASA and the future of the human spaceflight program is mired in partisan divides in Congress. In the next chapter we will describe this initiative more and see how it may yet provide the sword to cut the Gordian knot of the inspiration vs. politics dichotomy.

CHAPTER NINE

Stepping-Stones to Mars

千里之行, 始於足下 *[Qian li zhi xing, shi yu zu xia.]*
A journey of a thousand miles begins with a single step.
—LAO TZU

Sending humans to Mars is a big step. It is beyond our current capabilities, and much needs to be developed in order to attempt it. It is too big a step to take all at once—both because of the cost commitment and the required number of years to make it happen. It is a bigger engineering step from Earth orbit to Mars than was the Apollo lunar landing from our non-existent spaceflight capabilities of 1960. This admittedly imprecise assertion is perhaps counterintuitive. But in the 1950s and 1960s, we were already developing big rockets and advancing supersonic piloted flight. The leap to Apollo was enormous, but it was fueled by a Cold War that provided technical, human, and financial resources. Today, sending humans to Mars is limited by three big things: no heavy-lift rocket, large and unsolved human life-support requirements for a three-year mission in a hostile radiation environment, and entry and descent of a large crewed vehicle into the ultrathin atmosphere of Mars. The engineering challenges are described more completely in the appendix. Unlike with Apollo, each of these tough technical requirements is unique for the Mars mission: we have no other need for heavy-lift rockets, long-duration, closed-loop human life support, or huge entry vehicles to survive in planetary atmospheres. The converse is true of the main humans-to-Mars competitor: robots. Robotics (and the associated electronics) are multiple-use technologies with applications on

Earth and in other space applications. It is not just because the human flight engineering challenges are big but also because two of them are still inadequately understood (the life support and atmospheric entry) that I feel that the step to Mars for humans today is bigger than the step to the Moon for humans was in 1960.

That is not to say we can't make the step. The lunar landing wasn't achieved in one step—it required a succession of Project Mercury, Gemini, and Apollo flights as steps to the Moon. Each one of those steps included exciting technology and created new human flight achievements. These sustained technical progress and, more importantly, engaged and sustained public interest. That is what we need for the steps to Mars: affordable technical and human achievements that interest the public. In this chapter a flexible path on which we place steps to Mars is described. It may be the only way to create a sustainable program—because it starts with modest and affordable missions that are, at the same time, new and exciting.

Getting from here to Mars is like crossing a river. The direct way to get there would be to build a bridge—but we have neither the materials nor the money now to afford this construction, and even if we started now it would take too long a time over which to sustain political support. In our case the bridge is composed of the parts I mentioned above: the heavy-lift launch vehicle and the long-duration crew support vehicle(s).[1] Our launchers can't reach beyond the Moon, and the crew support vehicles have a life-support system limited to just a few weeks—nowhere near enough to support the multiyear Mars trip requirement.[2] To deal with the conundrum in human spaceflight (too much ambition, too little money), a high-level Review of U.S. Human Space Flight Plans Committee was chartered by the new Obama administration in 2009 to consider how a sustainable and worthy human spaceflight goal might be achieved. The committee, chaired by the former chairman of the Lockheed-Martin Corporation, Norman Augustine, introduced the idea of taking a *flexible path* to extend human flight beyond the Moon and toward Mars, without building a costly lunar-base infrastructure. The idea was to prepare for long-duration flight before facing the challenge of building the landing infrastructure. The administration embraced the flexible path idea, cancelling the (then-unsustainable) Constellation Program and setting new human spaceflight goals to reach a near-Earth asteroid (NEA) by 2025 and move on to flights in the 2030s to at least the vicinity of Mars. Landings on the Moon and the development of heavy-lift rocket capability for humans were postponed for several years while technology to accomplish the longer-duration, deep space missions was developed. In the interim, NASA would nurture a commercial launch industry to carry out low Earth orbit missions, supplanting the shuttle fleet and supporting the International Space Station.

Returning to the river-crossing analogy, the new approach was to cross the river on stepping-stones conveniently located so that we reach each step after achieving the earlier one. Near Earth asteroids form the first set of stepping-stones; the Martian moons, Phobos and Deimos, form the last set. The idea was to advance human spaceflight capability step-by-step, each step going farther and traveling longer into the solar system—with progress determined by going as you pay (i.e., with steps affordable within government space budgets). Each of these steps, like the Gemini and pre-lunar Apollo flights, should also be publicly engaging—thus generating, not diminishing, political support over the long time required for them.

The Augustine Committee wanted both a sustainable program (politically and economically) and one that was "worthy of a great nation"—that is, one where each step on the path was a significant new achievement advancing science and technology and bringing us closer to Mars. The administration embraced the flexible path concept, but even the first step on the path was too big—the requirement to send humans to a NEA by 2025 exceeded the capabilities that could be expected in the new (congressionally mandated) SLS/*Orion* program.[3] Only one suitable asteroid was known to be reachable, and even that object required greater launch energies and longer trip times than either SLS or *Orion* could support. The mission might be mounted later as SLS and *Orion* performance improved, but that would greatly delay the date when astronauts could journey beyond the International Space Station. The SLS/*Orion* could reach lunar orbit or an Earth-Moon Lagrange point (see figure 21), but these would be unsatisfying destinations, empty space, offering little exploration challenge for its astronaut crews. It was thought, for awhile, that the first milestone beyond the Moon could be a "Lagrange point" in the Sun-Earth system. There are five such points of gravitational balance between the Earth and Sun where spacecraft can more or less hover in their gravitational fields. Three of the points lie in a straight line connecting the Earth and Sun (L1 is near Earth on the side toward the Sun, L2 is also near Earth but on the opposite side, and L3 is on the opposite side of the Sun). The other two L points lie in the orbit of Earth around the Sun, 60 degrees behind Earth (L4) and 60 degrees ahead of Earth (L5). These are mathematical milestones—nothing is known to actually be at any of them (although there may be asteroids too small to have yet been detected trapped in the gravitationally stable regions[4]). Sun-Earth L1 is of particular interest for solar weather monitoring—observing the Sun's coronal and sunspot emissions to give a few minutes' advance

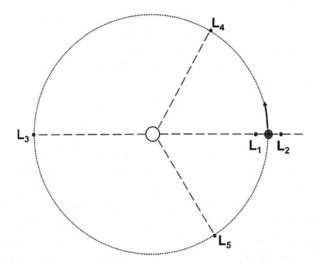

Figure 21. Lagrange points.

warning of solar storms that can disrupt power and communications grids on Earth and satellites in orbit around Earth. The United States has kept a solar weather monitoring satellite there for many years.

However, even a mission to nearby L1 (about 1.5 million kilometers from Earth toward the Sun) would take 2–3 months—beyond the capability of SLS/*Orion* for at least another decade. If only there was a nearer asteroid just beyond the Moon that we could reach—it could be the first step on the flexible path. But there isn't, and so clever space scientists came up with an idea: if humans can't go to an asteroid, then we'll bring an asteroid to within the reach of humans. Thus was born the concept of asteroid retrieval. Since we can't reach the first stepping-stone in the river symbolizing our way to Mars, we'll pick one up on the riverbank and toss it into the river and then start the human journey.

The idea to robotically move a celestial body may seem daunting at first, but several studies had previously shown that moving something approaching a thousand metric tons (approximately the mass of the International Space Station) was feasible with solar electric propulsion. Electric propulsion systems generate small but continuous thrust, and over long-enough flight times, they can achieve the velocity change (delta-v) required to move the asteroid to a new orbit. In 2009, JPL conducted a study, led by their chief electric propulsion engineer, John Brophy, to examine the requirements for an electric propulsion system that could move an asteroid into Earth's gravity well, to as low an orbit as the International Space Station.[5] Around that same time, a Planetary Society student intern, Marco Tantardini from Italy, informally promoted the idea that moving an asteroid closer to Earth would make it more accessible for commercial exploitation of its putative resources.

In November 2010 a group of us, including Brophy and Tantardini, proposed a study to the Keck Institute for Space Studies (KISS) at Caltech to consider the question posed at the beginning of the previous paragraph, specifically whether a reasonably sized NEA (e.g., with a diameter larger than 5 meters) could be moved to a point to serve as an exploratory target for a first human mission beyond the Moon. The asteroid could be both a "natural" space station (without the cost of building a new one beyond the Moon) and an object of serious scientific study. Near-Earth asteroids were already targets for robotic missions and increasingly of interest to those who saw them as a source of valuable space resources—propellants for further solar system missions or chemicals and minerals for commercial space applications.[6] Hands-on investigations by an astronaut crew would develop human space operations capability and could return important new information about the nature, structure, and composition of asteroids.

It was found that a solar electric propulsion system (SEP) of 40 kilowatts (end-of-mission power level) could move an asteroid weighing less than one thousand metric tons to the vicinity of the Moon in about seven years, providing the required velocity change. The mission duration could be shorter if the SEP spacecraft could be launched directly onto an escape trajectory from Earth and not have to spiral out under ion thrust from low Earth orbit. Two options for capturing an asteroid were identified: (1) capturing a whole one with an inflatable bag enveloping the 5–10 meter object, and (2) capture a boulder on a bigger, better-known asteroid with a robotic retrieval grabbing mechanism.[7] Later (after several years of follow-up studies), NASA selected the second option for the mission, bowing to the advantages of selecting a known target.

Once in the vicinity of the Moon, lunar swingbys could be used to modify the orbit so that the spacecraft could be placed at the Earth-Moon Lagrange point or in a very large orbit around the Moon extending even farther than the L1 and L2 points. This lunar distant retrograde orbit (DRO) was chosen to guarantee orbit stability: even if spacecraft control were eventually lost, the asteroid would impact the Moon far into the future (more than 200 years), eliminating any threat to Earth. The lunar orbit destination also provides two important exploration advantages: the human exploration target would be the first beyond the Moon, extending human

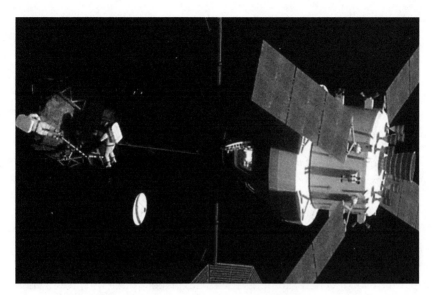

Figure 22. Asteroid visit by crew to retrieved asteroid. NASA.

flight to greater distances and flight time outside the Earth's radiation belts, and the final orbit of the asteroid would be attractive for observing the Moon and supporting eventual lunar missions. Both advantages further NASA's Human Exploration Program by developing capabilities for human Mars exploration.

A captured asteroid in lunar orbit could be reached by astronauts in an *Orion* crew vehicle launched on the SLS. To deliver the asteroid by 2025, the robotic mission would likely have to launch by the end of 2020. If the astronaut visit takes place by 2025, then the Obama asteroid objective, part of the new flexible path approach, will have been met. In today's highly constrained budget environment, this is the only way for humans to reach an asteroid (or land on any celestial object) by 2025. That accomplishment offers a promising first step in opening up the solar system to a new generation of human explorers, exceeding Apollo's reach and bringing new worlds and new ventures within our grasp.

The Asteroid Redirect Mission (ARM) provides an exciting human exploration opportunity that advances NASA's ability to reach other deep space destinations: more distant NEAs, the lunar surface, and the Mars system. The mission is an affordable way to build deep space operations experience and prove critical technologies like SEP and the SLS/*Orion*, while at the same time putting a high-value asteroid target within near-term reach of astronauts. Thus, it meets the two Augustine Committee criteria of sustainability and "worthy of a great nation."

Although a pair of robotic asteroid sampling missions, *Hayabusa* 2 (Japanese) and *OSIRIS-REx* (American), may succeed in returning small samples (less than 100 grams) from NEAs by the early 2020s, a visit to the redirected asteroid will enable trained scientist-astronauts to select hundreds of kilograms of samples (surface and interior) from an unexplored class of small bodies. The initial *Orion* crew will open the capture envelope to access several surface locations and extract core samples penetrating at least several tens of centimeters into the asteroid. These will be returned in *Orion* for detailed analysis in terrestrial labs.

The astronaut crew should also emplace long-lived science and planetary defense payloads on the NEA. These investigations may sound the asteroid's interior structure, measure its interior thermal profile, characterize the surface's optical properties, and assess the object's cohesiveness, porosity, and mechanical strength. A useful planetary defense investigation will be to obtain a value for the momentum multiplier value critical to understanding how to deflect an asteroid by kinetic impact. Other possibilities include measuring the ability of a spacecraft to tug an asteroid or even trying to deflect one with a dedicated ion beam from a solar-electric engine.

Opening a new era of space resource use, astronaut-returned samples could be used in experiments to extract water, other volatile compounds, and valuable metals (a C-type asteroid might contain as much as 10 to 20 percent water). Following initial astronaut visits, NASA should invite commercial firms and international partners to join it in using the captured NEA as a test-bed for mining and extraction technologies. Thus, ARM will enable concrete steps toward in-space propellant production (water into hydrogen and oxygen), a key to solving the logistics needs of wide-ranging human exploration. Before leaving the NEA, the first *Orion* crew should outfit the shepherding ARM spacecraft and the asteroid surface with navigation, grappling, and anchoring aids that pave the way for follow-up visits by robotic and human explorers.

Placing the asteroid in lunar orbit is not so much a destination or goal for human spaceflight as a technical milestone leading to the real goals at the Moon and Mars. The asteroid in-place could serve as a relay station or temporary base for tele-operated control of lunar vehicles and for lunar observations. Once the technology of asteroid retrieval is in hand and the observation program succeeds at finding many accessible small asteroids, future missions of more distant milestones such as Sun-Earth Lagrange (SEL) points can be designed. A Sun-Earth Lagrange point could be the next step outward on the flexible path into the solar system—it is truly in interplanetary space, and it might be accomplished with mission durations of about three months. In a subsequent KISS study, we identified other trajectories that might be used for interim missions on the way to Mars: orbits resonant with Earth, so that that once on them the astronaut returns to Earth. With asteroid redirect technology we could place an asteroid on a six-month or one-year resonant orbit with Earth. Such missions await an upgraded SLS booster or other heavy-lifters combined with orbital rendez-vous. This step might be succeeded by expeditions to reach more distant asteroid orbits, with mission durations of up to one year, or even all the way to Mars, for example to rendezvous with one of its moons (Phobos or Deimos). Readiness for these missions will be determined by the capabili-ties and availability of heavy-lift launch, crew life support, and deep space habitats. Ideally, the deep space interplanetary expeditions will be under-taken in the late 2020s and 2030s, but if funding falls short, or heavy-lift and crew capsule development slip, then multiple astronaut missions to the captured asteroid in lunar orbit can continue to develop human spaceflight operations experience and resource extraction processes.

One of the greatest challenges for human interplanetary space flight is protection from cosmic rays and solar particle outbursts. The only current protection method is shielding, although biological and even nutritional research continues on mitigating radiation. Shielding takes mass. An aster-oid has mass. Shielding mass could be a most useful asteroid resource, and our ability to capture and tow asteroids may furnish vital natural shielding for astronauts in interplanetary space. Nathan Strange, who co-led the KISS follow-up technology workshop with John Brophy[8] and me, has suggested that an asteroid-shielded habitat module could be put on an Earth-Mars "cycler" trajectory discovered and analyzed originally by Apollo astronaut Buzz Aldrin. The cycler trajectories connect Earth and Mars and can, in theory, be used repeatedly, albeit with a careful hyperbolic orbit rendez-vous, for regular human missions to Mars. Another possibility for asteroids in a human space transportation system is as a source of propellant. Up to 10 or 20 percent of a carbonaceous asteroid's mass may be water: if in situ resource utilization systems can extract that water (via solar heating), it may provide the great bulk of propellants necessary for interplanetary flight.

In these ways, asteroid redirect technology becomes much more than a simple one-shot mission enabling humans to advance beyond the Moon much sooner than would be possible if we had to wait for new rockets and crew vehicles to reach deep space. It creates an architecture that permits deeper and deeper steps into the solar system, a platform for astronaut experience and operations, and perhaps even the basis of a transportation network for safe long-duration human spaceflight.

Many in the space community don't like the stepping-stone approach created by using near-Earth asteroids to go into the solar system. Space enthusiasts are focused and destination oriented. Some want to "go back to the Moon, this time to stay"[9] (i.e., creating a lunar base), and some want to go directly to Mars. (Many of course want both—back to the Moon and on to Mars.[10]) I am a space enthusiast, and the Mars destination is much more appealing to me than is step-by-step on a flexible path. But we have now proved several times that such enthusiasm doesn't extend to political (or even engineering) reality. Endlessly having the wish foster the thought of a big space race back to the Moon[11] or Mars has kept us (all nations) confined in Earth orbit for four (soon to be five) decades, and as attention turns to other areas of space achievement (such as robots for interplanetary flight or tourists to space hotels), it is likely to keep us here forever.

Imagine, however, the stepping-stone approach. The first step would be a human going beyond the Moon in (say) 2025, getting to and bringing back pieces of a near-Earth asteroid that has been moved there in a remarkable robotic adventure (say from 2020–2024)—altering the course of celestial bodies for the benefit of humankind. This step is achievable in today's space program with little or no increase in space budgets. Then a second step—either doing more with longer excursions at that near-Earth asteroid or going to one farther out, beyond the Earth's sphere of influence into interplanetary space. Doing this with human flights from 2025–2028 could and would bring us to a first interplanetary gravitational milestone, the Lagrange point where Sun and Earth forces balance or on longer missions on an Earth-resonant orbit. We could move another near-Earth asteroid there and make it both a base for solar weather monitoring, helping to protect our Earthly technological assets from radiation storms, and a shield for human habitat in space, protecting the astronauts. By this time I would hope we would take the heavy-lift rocket development step and be building crew support vehicles for multi-month flights in space.

The next step might be onto a cycler trajectory or a Mars round-trip mission using mass from a redirected asteroid we placed in an intermediate orbit. This is the advantage of the "flexible path"—to make the next step selection based on how we achieve and what we learn from the previous step. This step might be possible in the five years after the previous one, e.g., 2030–2035. After this, we would take the leap to Mars—likely first to Phobos or Deimos, which are basically near-Mars asteroids, and thus our experience with asteroids is applicable to them. Human landings on

Phobos or Deimos are much easier than landing on Mars (they have no atmosphere and essentially no gravity, making the landing more of a simpler rendezvous problem, like that of an asteroid). From the Martian moon humans could conduct detailed reconnaissance of candidate landing sites on Mars, tele-operate robotic precursors, and observe tests of the eventual human landing vehicles. Then we would be ready for the landing on Mars—by 2040 or so.

As noted earlier, entry and landing on Mars is an enormous challenge, one that will require large and complex vehicles—which in turn will require expensive testing. Likely there will be testing on Earth with re-entry simulations from low Earth orbit or in atmospheric flights: on the Moon at least for the propulsion and landing systems and on Mars with robotic missions prior to the human-crewed mission. The cost of the entry, landing, and descent system is the part of overall system development that we will want to defer until after we have built the rockets, interplanetary vehicles, and life-support systems that are necessary. Scheduling the landing system development in the time scale of the above scenario means doing it in the mid- to late 2030s—both for the lunar landings and the robotic Mars landings. Again, that will disappoint space enthusiasts who argue for a return to the Moon much sooner. But seeking to incur that cost much sooner is what has proved to stall human spaceflight in Earth orbit. Consider that even in the George W. Bush administration with a commitment to return to the Moon (and a NASA administrator dedicated to lunar base development), the planned lunar landing had slipped to the end of the 2020s decade before even a single piece of flight hardware had been developed. It would have slipped more—which is why the Augustine Committee found the program unsustainable. NASA was incurring costs for a lander development 15–20 years before it was needed, and those costs competed with other necessary developments—which brought the program to a standstill.

Any large human space endeavor will be conducted both for reasons of national interest and with international cooperation. The humans-to-Mars endeavor certainly will. This pattern has been nicely set with the International Space Station. We discussed why in the last chapter. But still nations (and maybe even private ventures) will participate with their own goals. There are many space races going on around the world for national prestige, technological leadership, economic development, and workforce and industrial engagement. While the United States pursues the flexible path as its approach to Mars with humans, Russia might build a new space station, and China and India might try a lunar landing and robotic Mars missions. At the same time they and others may collaborate internationally

Table 6. Potential human spaceflight achievements

Year	Achievement	Led by
2018–2024	Retrieve NEA into lunar orbit	United States
2024–2026	Human mission(s) to retrieved asteroid	United States with others
2016–2024	Robotic lunar landers	Russia, China, India, maybe Europe
2017 . . .	Robotic Mars landers and rovers	All major spacefaring countries
2019 . . .	Commercial human flight in Earth orbit	Private
2024–2030	Asteroid retrieval to Sun-Earth L1 or an Earth resonant orbit	United States with others
2024 . . .	Commercial robotic lunar and asteroid activity	Private
2026–2032	Human lunar landing	China and Russia with others
2026–2032	Humans to L1 or to Earth resonant orbit or to asteroid in natural orbit	United States with others
2030–2034	Robotic precursors for human Mars landing	United States with all spacefaring nations
2034–2040	Human Mars flight-orbiter (no landing), Phobos landing	United States with all spacefaring nations
2040–2049	Human Mars landing	United States with all spacefaring nations

on other activities in space. I have frequently been asked, "What if China gets to the Moon before we do?" as if this is some kind of threat to our security.[12] I have two answers: (1) they can't, the United States already got there first, and (2) what does it matter—if the United States is doing even greater things moving out into the solar system, conducting experiments on asteroids, and traveling toward Mars, their presence on the Moon will be welcome. The United States could even help them or be part of an international lunar base led by China, Europe, and Russia. So long as the

United States is doing its own great things, then everyone benefits and cooperation can proceed broadly.[13]

Let's summarize this chapter and our steps to Mars in table 6. It is just a scenario, not a prediction and not a plan—what NASA likes to call notional, and I call illustrative. It is intended to show the reader that what we lay out is a credible and affordable approach to human exploration that is politically realizable in many nations. Most of all it would be publicly exciting to the whole world—with new achievements for some and benefits to all. The details can, and would, be changed—but this outlines a credible approach for revitalizing human spaceflight.[14]

CHAPTER TEN

Implications and Conclusions

We shall not cease from exploration, and the end of
all our exploring will be to arrive where we started
and know the place for the first time.
—T. S. ELIOT

One of the reasons I love working in the space program is that it allows me to think about really deep questions of life and our universe — not just with thoughts, but also with action — going after the data that illuminate them. That process of going after the data provides exciting and interesting opportunities to work on great, one might say epic, adventures. Deep questions are those like, Are we alone? How did life originate? What is its destiny? Is life ubiquitous or unique in the universe? Is its evolution into intelligence an accident or is it inevitable? etc. For all of history humans have addressed these questions. In myths, folklore, stories, religion, philosophy, and pseudo-science, we have spent enormous amounts of time and energy (and huge shares of cultural and national wealth) addressing these questions and seeking answers. But, in my lifetime, we could finally address them with exploration and science — by experimentation, measurements, and travel to other worlds. And the share of national wealth it requires is pretty small — less than 1 percent of the American federal budget goes to space exploration (and for other nations it is even less).

In addition to the questions posed above about the origin, evolution, and nature of life, we also have existential questions to ponder, such as the likelihood of planetary catastrophe and species extinction. I am pretty sure humans are the only species on Earth with self-awareness questions about life. If you agree with my argument in chapter 6 about the unlikeliness of extraterrestrial intelligence, we may be the only species, period, that ponders such questions.

These questions drive space exploration, which makes being a part of it so exciting. I have a simple definition of exploration: it is discovery plus adventure. In space we discover new fundamental things about ourselves and our universe, and we do so by conducting high-risk, creative adventures. And that brings us to the implication and conclusion of this book: the future for humans in space is to become a multi-planet species with no existential threat,[1] living physically on two planets and exploring the universe mentally on many others, learning about life within ourselves and in the universe.

In this optimistic future,

1. human presence will extend indefinitely beyond the solar system with small spacecraft and evolving robotic and information processing technologies. Exploration of other worlds will be "virtual," evolving with the human species and our technologies. It will be as interactive and humanly satisfying to those in future centuries as physically being there was to those in past centuries—maybe even more so given the evolution of biological and molecular technologies and the speed of our adaptation to them. We will, however, send our human form no farther than Mars.
2. we must, however, send humans to Mars; otherwise, we are doomed to live within both a physical and psychological limit on a fragile planet. Mars is great: it is (relatively) nearby with the stuff (water, oxygen, soil, gravity) to support life and a land area equal to that of Earth. It is the only accessible, habitable world beyond Earth. But, Mars settlement is a huge question—one that will take centuries to determine.

One might argue that the first conclusion could apply also to Mars—we can virtually explore it as well. But I don't think so, if for no other reason than as a back-up to Earth, humans will go and possibly even settle on Mars, sometime in the next few centuries. Furthermore, the time scale for Mars is very different than for any more distant human space activity. Mars can be reached in just a few decades, and we could (indeed should) have dozens of missions there in this century. But reaching beyond Mars—even if we wanted to—is not possible in this century, and as noted, by then the technology and mind-set of exploration will have moved on.

Mars as a physical limit for humans is not much of a limit. With a land area equal to that of Earth's, it will take us decades and more likely centuries to explore, centuries or perhaps a millennium to settle, and a millennium or more to then terraform. That is long enough for planning at least, without going into science fiction. It is also long enough to keep humans in space both physically and intellectually occupied and to satisfy all our wanderlust.

These conclusions have implications for space programs around the world. Sen. William Proxmire[2] (and other politicians) famously asked us (during congressional testimony by The Planetary Society), "Mars has been there for billions of years; do we have to rush to get there now?"

Our answer in the past was, "No, we don't have to get there now," but going there isn't about Mars—it is about us—who we are and what we want to be. Our civilization and our society will be better for using long-range vision to guide wise scientific, technical, and cultural accomplishments, and this includes creating the first steps of humans into the solar system on to Mars. But the answer is changing; now we might say, "Yes, we do have to rush (or at least hurry)." In addition to the sending of humans to Mars being about us, it is also about humankind and what it means to be human. Right now, we have yet to reach full cultural acceptance and satisfaction with the virtual world. Public interest in space exploration is still dominated by human interest. But it won't be for long. If another decade (let alone two or three) goes by without humans going farther in space, not only will public interest likely decline but also the aforementioned robotic and virtual exploration technologies will pass us by. We would be hidebound on Earth—a very unsatisfying end to humankind's growth.

Vision can guide, but "money is the mother's milk of politics" and sending humans to Mars is a practical budget issue—for the United States and other countries and for putative private or commercial interests. In the previous chapter we described the difficulty of getting a political decision to send humans to Mars—it is too big a political step, both in money and time. The political conundrum is that Mars is too big a political step but that giving up on it is politically unacceptable. No other space destination goal makes sense for humans, and no spacefaring nation's leader can cede his or her country's interest in it. The U.S. president cannot say, "OK, let's give up on human spaceflight; that was so twentieth century; we'll step back and let others (Russia, China, India, etc.) do it instead." The Chinese premier cannot say, "We'll let other countries go to space; we'll stay back on Earth." Russia space policy asserts human exploration of Mars as its goal, and Europe, Japan, and India have burgeoning exploration ambitions already reaching Mars, as do even less industrialized countries. Human spaceflight has always been conducted for geopolitical positioning—not for science or the economy. In some ways it seems like a rite of passage for nations transitioning into a technological and global future. I am deeply impressed by the Chinese and Indian space programs reaching out with human spaceflight and planetary exploration and by the resurgence of the Russian space program funding and developing a spate of plans for lunar and Mars missions and new rockets. I think it is not lost on the leaders of that nation that in the long course of history the Soviet Union will be remembered with only two great (positive) accomplishments: victory in the Second World War and human spaceflight.

The current U.S. administration has set the goal of achieving human spaceflight to Mars in the 2040s;[3] Russian leaders have talked about achieving it on a similar, if not faster, time scale, and other space-faring nations have documents stating it more vaguely as a goal in future decades. Curmudgeons in the space business point out that since the beginning of the space age the goal has always been 30 years in the future and it is likely to remain so. We might worry that advances in robotic exploration and capabilities will lessen the interest in human missions or aid arguments against spending money on human flight. I think the reverse will happen—as robotic missions accomplish more, society will be even more compelled by the lure of Mars and unwilling to cede it to the robots. At least for this century, I think we will remain on the side of the humans (although I admit to the danger of the wish fostering my thought). In part this is cultural—we still will be motivated to do what we can do, and we can reach Mars with humans (as opposed to reaching Alpha Centauri, Pluto, or even Europa). The success of, and the public interest in, the robotic Mars rovers suggest we are inexorably progressing to human exploration there. For at least the next half century, humans would be able to explore tens to hundreds of times the area on Mars in a given amount of time, compared to robot rovers. More than area, they will do much more—scientifically building up the knowledge and eventually the infrastructure to determine the future of human habitability. Mars exploration begets more Mars exploration, and that is why I think the goal of 30 years in the future for humans is real now.

Still, translating the political inevitability of humans to Mars into political action is the political process—and it is messy. The cost of human spaceflight is the real inhibitor, although the money spent is all spent on Earth by citizens who get the benefits of jobs, technology, science, education, and joy from space exploration. This is why the third world nations pursue space adventures and why the industrial nations can't afford to give it up. The Congress of the United States has many times shown it is willing to spend money for these benefits even when the projects are unnecessary or without other benefit. The space shuttle and the International Space Station were great technical achievements (among the greatest in history)—that they were neither necessary nor had science or consumer benefit is pretty much forgotten and maybe, in the grand scheme of things, irrelevant. Sure, we could have done better—in some ideal world we could have advanced space exploration in much better ways and perhaps even be on our way to Mars now. The shuttle and the International Space Station lacked some positives, but they had no negatives—and that's pretty good in today's political environment.

We could be repeating that experience now in the decade of the 2010s—building a rocket we don't need to compete with a world supply of rockets we already have and that are underutilized. It won't be bad; it just isn't as good as it could be. Human spaceflight is expensive and risky, and it should be things that are worth the high cost and risk, not just routine flights to low Earth orbit. Humans should be exploring—going faster and farther into the solar system and toward Mars.

My point is that money is not really the big issue: no one is going to discontinue human spaceflight. We are going to spend money on it, either for worthy goals (exploring other worlds) or mundane ones (trapped in Earth orbit with contrived science experiments). But, the worthy goal is a long-range goal: not to be achieved within a decade like Apollo under the constant pressure of a Cold War but a several-decades-long goal with fits and starts and budget cycles and problems that inevitably stretch the goal long and political support thin.[4]

This situation defines the problem: keep the public engaged with exciting milestones while making steady progress to the ultimate human-to-Mars goal. Apollo did this with great flight milestones in Mercury, Gemini, and the Apollo flight prior to the *Apollo 11* moon landing. Now human spaceflight needs to do something similar: achieve further, faster, and longer flight goals in a steady stream of progress between Earth and Mars.

This is why the Augustine Committee (chapter 9) recommended the flexible path in order to create both a "*sustainable*" program and one "*worthy*" of a great nation." Steps on that path would start with flybys beyond the Moon, lead into interplanetary space with milestones like Lagrange points (of Earth-Sun gravitational stability) and landings on near-Earth asteroids farther out in interplanetary space, and then progress perhaps to Phobos or Deimos around Mars and eventually to Mars. It is not hard to see that such milestones, accomplished each couple of years, breaking speed and distance records and new life-support achievements, would engage the public and keep momentum going for human spaceflight.

The first step is the hardest (at least for building momentum). Doing just a flyby past the Moon seems kind of empty without a physical destination. We could build a space station beyond the Moon, but that would be a long and costly diversion. We could fly round trips to the point of Earth-Moon gravitational stability, but those are mathematical destinations not physical ones for exploration. We could attempt to reach the nearest of the near-Earth asteroids, but the mission requirements even for that are in excess of what we can currently afford to build in the NASA or other nations' space program. We may wrestle our way out of all these difficulties, but I strongly submit that there will be no more exciting or technologically beneficial way to do so than by robotically rearranging a tiny piece of the solar system to bring it a bit closer—to retrieve a small near-Earth asteroid (or even a piece of one) and make it the goal for the first human step beyond the Moon. The Asteroid Redirect Mission is described in chapter 9.

Thus, we solve the dilemma of reconciling the cultural inevitability of human spaceflight to Mars with the political and economic difficulty of getting it approved in practical and sustainable government budgets. We also harness creativity in both robotic and human spaceflight to provide a cultural synergy for robots and humans working together. This fits with the book's broader theme, viz. that extending the human presence and indeed the human destiny beyond our own planet is a human *and* robotic activity. By getting rid of the shibboleth of "Moon, Mars and Beyond," our human spaceflight program can rid itself of the implicit requirements of the "beyond." The program will be more realistic, hence more credible and hence more politically supported. At the same time it is more exciting with an existential destination. Humans can go to Mars to start being a multi-planet species and then from their Earth/Mars home extend exploration and themselves to the outer solar system and to interstellar space through evolving technologies of robotics, information processing, molecular engineering, and biogenetic sensing.

We will be exploring virtually, but our presence will be extended by interacting even with (putative) life "out there." Extending human presence to the stars and even to the outer parts of our own solar system (beyond the main asteroid belt to the environs of Jupiter, Saturn, Uranus, Neptune, and the Kuiper Belt) will be done without human flight. At the pace we are advancing human spaceflight, we will be very happy to get Europa and Titan robotic surface explorers in that time. And by 2100 those robotic technologies will be re-creating Europa and Titan (for example) in virtual worlds for us to explore here on Earth. By then, too, we'll be on our extended habitat on Mars. Home by this time will be Earth and Mars (that part is relatively easy to grasp); the harder-to-grasp part is how we will *feel* about our "couch-potato" status, half exploring and inhabiting other worlds without leaving our home. I said how *we* feel—but it is not "we" who are relevant. It is our descendents, a couple or more generations after us. My guess is they'll be fine—they will be a multi-planet species with humans on Mars and human presence beyond.

Appendix

A Human Mars Mission

*The Earth is the cradle of humanity, but mankind cannot
stay in the cradle forever.*
—KONSTANTIN TSIOLKOVSKY

This appendix explains more about the engineering and other challenges
for human Mars flight. The challenges are big, and although I am unabash-
edly in favor of human exploration of Mars and I think it is absolutely
inevitable and necessary that humans will go to Mars, we still must be
realistic. We are decades away (as described in chapters 8 and 9) from a
first human Mars mission, but more importantly, it will take centuries (if
not millennia) after we first get there to explore and maybe settle there.
Becoming a multi-planet species on Mars is a human task for centuries.
Beyond Mars, during those centuries humans will explore other worlds
virtually (as discussed in chapters 7 and 10).

In chapter 9, I suggested that performing a human (crewed) mission to
Mars now was a bigger engineering challenge than reaching for the Moon
in the 1960s, even though then we[1] were starting without even a space
program. That is because then we had the basic elements of large rocket
development and crew support underway—through our military rocket
and aircraft programs. Today, of course, we have much more underway in
a rich and multifaceted space program, in the United States and in other
space-faring countries. Robotic spacecraft have explored all the planets
in our solar system and dozens of moons and asteroids and landed on the
Moon, Mars, Venus, and a near-Earth asteroid. We have built dozens of
rockets to launch machines from Earth and several to launch people. We

developed a reusable shuttle and built an incredible space station. Indeed, we have much more than we had in the 1960s. But we also have much less. We have no heavy-lift launch vehicle—the United States abandoned the Saturn V rocket and the Russians abandoned the Energia rocket. Our crew transport is limited to just days or weeks of life support. The space suits that crews use for extravehicular activity are basically those of the 1960s—no more flexible and no more suited for planetary exploration now than they were then.

Prior to the human Moon missions, there were environmental uncertainties: some worried about lunar dust swallowing up or contaminating anything that landed there, and quarantine of astronauts coming back from the Moon was initially deemed prudent to guard against putative space bug. Those concerns dissipated quickly. But those environmental uncertainties were miniscule compared to those of a human mission to Mars. For starters, we can never prove there is not life on Mars (because you can't prove a negative; the best you can do is say it is unlikely). Thus, sending humans will require extraordinary care concerning possible unknown interactions with possible unknown life. Then there is radiation: the surface of Mars is one of the best places available to us for sterilization. There is little protective atmosphere there, and thus solar and cosmic ray radiation falls unhampered on the surface—lethal for any life-form and creating toxic interactions with surface materials. Another environmental uncertainty for humans to Mars is the effect of either weightlessness or centrifugally created artificial gravity during the multiple years of space travel. None of these are showstoppers: only the grumpiest of naysayers think these problems are unsolvable. But they are all tough, and it will be expensive and take many years to deal with them—much more so than dealing with those of the human lunar missions.

Many space vehicles are required for a single human mission to Mars (let alone for continued exploration). How many depends on many assumptions—some technical, some financial, and most political. These are my assumptions:

1. It (the first mission) will be a multinational, internationally cooperative mission carried out by government space agencies (hopefully with private and even some commercial participation).
2. It will be round-trip. (A private organization has suggested a one-way trip to Mars, and that indeed may follow—just as the Jamestown colonists came to Virginia on a one-way trip 117 years after Columbus's round-trip mission to the Americas. I estimate the 117-year time

period between the first private one-way settlement attempt and the first government round-trip mission to Mars to be about right.)

3. It will be to the Martian surface (likely following one or more human missions to the vicinity of Mars—e.g., circumnavigating (as proposed by another private company) and/or orbiting to Phobos or Deimos.
4. It will involve a crew of four to six (likely six), at least three of whom will go to the surface.
5. It will be carried out with the technologies of the 2020s (consistent with a mission to be accomplished in the 2040s).

I also assume that while the mission will be multinational, the official language will be English and the standards will all be metric. All of these assumptions have been discussed in the book; they might be considered conservative, i.e., no crash program of political support, no private venture for glory or commerce, and no technology breakthroughs. However conservative, they are more optimistic than are perhaps warranted considering the current trend in space budgets.

The vehicles required for sending humans to Mars, supporting them, and providing for their return to Earth are as follows:

* A crew module for the interplanetary flight (habitation module)
* An entry vehicle for Mars—probably something shaped more like the space shuttle than like the entry capsules used on robotic missions. The entry vehicle will rely far more on propulsion and descent engines (as opposed to the shuttle, which relied on aerodynamic deceleration and control) since we will have to be able to control it for pinpoint landings and will have to do it in a very thin Mars atmosphere. In addition to the entry vehicle meeting descent and landing requirements, it may also be used for aero-braking or aero-capture-assisted entry—that is where the drag of the atmosphere is used to slow the vehicle down from orbital to entry speeds. The drag from the thin atmosphere will be small, but we will still have to deal with it and the heating from its friction with the entry vehicle.
* A crew (habitation) module for the Martian surface—similar to the interplanetary crew module. How one splits up the crew (all or just some of them on the surface) and the mission time (how many days on the surface) has to be decided. The relative geometry of Earth and Mars orbits constrains the mission times both for interplanetary transfers (from and to Mars) and for the stay time at Mars. The former are approximately 6–7 months, and the latter are approximately

18–21 months.[2] Not all of the stay time at Mars will be on the Martian surface, although exploration time there is obviously more desired than boring waits in Martian orbit for Earth-Mars geometry to align.

* A rocket for the crew to get off Mars. The ascent stage will be designed to reach Mars orbit where it can rendezvous with an Earth-return vehicle. Fuel for the rocket and fuel for powering the surface habitat of the crew can be carried to Mars from Earth, or it can be manufactured using indigenous Martian resources. These include carbon dioxide and water in the atmosphere and perhaps accessible subsurface water ice. These can supply the basic hydrogen, carbon, and oxygen for making fuel.

* Support cargo for Martian surface operations—including that in situ propellant production plant just mentioned, as well as other consumables for life support of the crew

* A rover for mobility for exploration on Mars by the crew. Perhaps also some smaller tele-operated robotic vehicles: a rover, a balloon, or a robotic helicopter or airplane might be used to extend the crew's exploration capability.

* A crew vehicle for Earth return, like that used on the way to Mars

* A capsule for Earth entry, possibly in Earth orbit waiting for a rendezvous with the Earth-return vehicle

* Propellant for all the above: trans-Mars injection; Mars orbit insertion; Mars entry, descent, and landing; ascent from Mars (unless made from in situ propellant production); trans-Earth injection from Mars; orbit insertion; and/or entry at Earth

* Robotic cargo carrying vehicles to support the infrastructure for the above; as suggested earlier, solar-electric propulsion may enable moving a large mass efficiently. The solar electric power system will have to be two to three times larger than for the ARM mission discussed in chapter 9.

Add all this up, and it totals more than 500 and perhaps as much as 1,000 tons to be delivered to low Earth orbit for a single human Mars mission. It's a whole different scale than Apollo. The most capable rocket in history, the Saturn V, could launch about 120 metric tons to low Earth orbit. The large Russian Energia booster would have launched close to, but less than, that (about 100 metric tons) if it had become operational. For comparison, the current largest booster on Earth (Delta IV) can only launch about 23 metric tons to low Earth orbit. The space shuttle was capable of launching about 20–25 tons, and versions of the French Ariane and

Japanese H-series rockets are planned with similar capability. The United States has embarked on building a new rocket, the Space Launch System, whose first version is targeted to be capable of launching 70 metric tons to low Earth orbit, while its Block II is supposed to be capable of launching 130 metric tons (as is the Chinese Long March 9, some years in the future). It seems highly unlikely (and very inefficient) that anyone will want to construct a rocket bigger than that—not just for technical reasons but also for economic ones. A super-heavy-lift human Mars rocket would have one purpose only—namely, transporting humans to Mars. The cost for a special-purpose, infrequently used, giant rocket is prohibitive—and remains an impediment to human Mars mission planning (in every country). Furthermore, it is not a smart way to design such a complex mission. Trying to plan a round-trip Mars mission on a single vehicle (even if we had one) would subject the designer to an extreme sensitivity to changes—a change of 100 kilograms (0.01–0.02%) required on the Martian surface requires about an extra 2,000–3,000 kilograms of mass required in low Earth orbit. That is not a robust design. Splitting up the launch requirements and then arranging the vehicles with the capability of bringing the crew and cargo together at Mars isn't easy, but it permits a much more robust mission design. And, unlike huge boosters, it has many applications—we can use multipurpose rockets and build an Earth orbit infrastructure, mixing and matching components for different missions. We can even split requirements and vehicles up among several nations and over a multiyear time schedule so as to make missions not just feasible but also (and this is key) affordable.

Thus, human missions to Mars will be designed with multiple launches—splitting up cargo and crew vehicles to fit on affordable, albeit still large, rockets. Over the years human Mars mission studies have led to designs requiring anywhere from four to ten launches for a single crewed mission to Mars. NASA's latest "design reference mission" (circa 2009) included at least eight heavy-lift launches for the mission, splitting up crew and cargo and "pre-deploying" some of the cargo two years prior to the arrival of the crew. An International Space University (summer student) project came up with a mission design requiring 35 launches (18 for cargo and 17 for crew, admittedly with large surface infrastructure development).

Apollo missions to the Moon were conducted with multiple vehicles on a single launch vehicle. Coordinating rendezvous between vehicles on multiple launches requires more coordination but not too much more complexity. One approach, frequently suggested but never yet implemented, is to create a "fuel depot" in Earth orbit that could be used to refuel vehicles after launch for the trip to Mars. The advantage of such a depot is that it could have multiple uses—for refueling Earth orbiters as well as interplanetary spacecraft. Multiple uses can lead to lower costs both by virtue of cost sharing and economies of scale. The disadvantage is that different missions have different orbits (not to mention different propulsion requirements), and a design of a single multipurpose fuel depot might be impossible or very inefficient.

As challenging (and expensive) as the launch of the required large mass and of all the necessary components of a human Mars mission is, it is, as noted, a challenge we know we can meet. The split of cargo and crew and deployment of multiple launches will involve many trade-offs—most of them falling into long-range vs. short-range decisions: planning multiple missions vs. planning a single mission. We may not yet have the vehicles, but we do have the know-how both to build them and assemble them.

A problem that we have not yet solved is handling the entry, descent, and landing for human-sized vehicles on Mars. The so-called "7 minutes of terror" we all watched when the *Spirit, Opportunity,* and *Curiosity* rovers landed on Mars pale into 7 minutes of parlor games in comparison to the real terror that will occur as astronauts in a human life-support mission vehicle hurtle through the Martian atmosphere hoping for a safe touchdown. Re-entry into Earth on the Apollo missions was a scary time—the astronauts hit the Earth's atmosphere with a speed of 11.2 km/s (~40,000 km/hr). The heat shield, necessary to protect the astronauts and their return capsule from the heat caused by friction with the Earth's atmosphere at this speed, was nearly 4 meters in diameter. That friction is what slowed the re-entry capsule down enough to permit safe parachute deployment and control of the landing of the capsule. For astronauts returning from Mars, the entry speed is going to be a bit higher (from the interplanetary trajectory), and thus the re-entry capsule will have to be heavier and stronger.

But it is the entry at Mars that is the new requirement. Mars's atmosphere is less than 1 percent that of Earth's, so it will require a much larger heat shield to slow the astronauts down, even at the lower entry speed of under 20,000 km/hr (coming out of Mars orbit, rather than directly entering the Mars atmosphere from an interplanetary trajectory). In addition, we are taking much more down to the Martian surface—a set of vehicles, power sources, life-support system, rovers, an ascent rocket for return, etc. These will be split up in multiple landers, but still the requirement will be to land 30 to 50 tons with each lander. Instead of an entry capsule, the entry and descent system will be more like the shroud used on top of rockets to carry payloads from launch to Earth orbit. A recent NASA study has concluded that parachutes may just not be suitable or feasible for landing humans on Mars. Even with the robotic Mars Science Laboratory mission, NASA had incorporated descent engines again into the landing system—together with the parachutes and aerodynamic deceleration. But for human-crewed missions the mass will be too great and the supersonic entry and descent will require a combination of propulsion and deployable rigid and/or inflatable aeroshells. Even though the Mars atmosphere is thin, it will be usful to lower the propulsion requirement to brake into Mars orbit. Dipping into the atmosphere can provide an "aero-assist" to the braking requirements. We will need all the help and control possible—bringing in the multiple, very large cargo payloads, not to mention the crew, to rendezvous on the surface is a huge task. I think accomplishing it is the biggest uncertainty in human Mars mission design—bigger even than the uncertainties of life support and protection. The comparatively small payloads for robotic missions (like the Mars rovers and even for a robotic Mars sample return) barely approximate the requirement. Larger ones will need to be tested on future missions.

Another big issue for the human Mars mission will be the in-space propulsion system (for trans-Mars and trans-Earth interplanetary flight out of and into orbits around Earth and Mars). Most human mission designers want to use a nuclear thermal rocket—that is, carrying a fission reactor onboard to heat the liquid hydrogen propellant. It is far more powerful than all chemical propellants. But it is a unique requirement—no other space development of nuclear rockets is likely. It will require years of expensive testing and development. An alternative just beginning to be studied is using solar electric power for the cargo vehicles (where long flight times at low thrust are acceptable). They can move large mass, and then only the lighter-weight crew vehicle will require the chemical propulsion. Solar electric propulsion is now extensively used in space (by many countries), and a large system has recently been proposed for the Asteroid Redirect Mission (ARM) described in chapter 9. If used there, it would become a component of the human spaceflight program and presumably more available for upgrade, to hundreds of kilowatts, into a Mars cargo carrier. Work on the ARM mission has led to Mars architecture that benefits from the large solar electric propulsion development.

Humans on Mars will need nuclear power. Indeed, machines, robots, and instruments will also need nuclear power. On Earth too we need nuclear power, but we cannot ignore that it is politically controversial and its development and employment are inhibited. Facing this issue sooner, rather than later, is better—unless we believe in the invention of technologies that are going to change dramatically the current debates. An example would be advances in solar cells that would make them much cheaper and more efficient or reduce their use of resources and consumption of land area, or conversely, new processing and storage techniques solving the problem of nuclear waste disposal. Although these are not space issues, solving them on Earth (or not) is important to their availability and cost in space applications. Although I agree that we will want to launch reactors for nuclear propulsion to Mars and nuclear power on Mars, I cannot predict that its use will be politically acceptable for the first human expedition to Mars. We should develop the solar electric propulsion and power alternatives.

As for propulsion to lift off Mars—we will carry along a rocket for the Mars ascent stage. We could carry along its fuel, but as noted above, it is expensive in terms of launch mass to carry added weight to the Martian surface (and propellant weighs a lot). Robotic precursor missions should establish the feasibility of making the fuel on Mars using Martian resources: hydrogen and oxygen present in atmospheric water vapor and subsurface water, carbon, and oxygen present in the atmospheric carbon dioxide. In fact, the human mission plan already consisting of predeployed cargo components on Mars will likely include an in situ propellant production plant, with all the fuel produced (and even loaded into propellant tanks) before the crew even leaves Earth or gets to Mars. An efficient in situ production plant will use nuclear power, although in principle large solar arrays (and they would be large) could be used.

Another large component of the exploration plan involves the surface habitat(s) and mobile vehicles for sojourns. Minimum energy missions dictate long stay times at Mars, and using that stay time for exploration on the planet is a prime mission objective. Stay times on the surface will be about 1.5 years. The requirements for life support and safety (not to mention comfort) are large, and the habitat module will require a great deal of power—systems and backup systems, and backups to the backups.

One big cost driver on the human Mars mission is likely to be "planetary protection" (although I expect the cost per word written about this subject will actually be quite small). Some people are worried about protecting Earth from putative Martian organisms that might be carried back by the crew—on themselves and/or their return vehicle and instruments. Remember the "quarantine" of the first Apollo astronauts brought back from the Moon—a bit ludicrous since they had splashed down in the ocean without any planetary protection. But ludicrous won't stop a public relations requirement, and even though the Martian surface is one of the most sterile places imaginable (because of its constant bombardment by ultraviolet radiation), there will still be calls for protection of Earth. True protection is expensive and will add to the mission cost. Perhaps a robotic precursor mission can be utilized to convince people there is no danger to Earth from the crew's return.

Planetary protection of Mars is another issue. Since we are looking for life on Mars, we surely do not want to confuse our search with Earthly life. Since Mars and Earth have already exchanged material, one might dismiss this concern. Indeed, some scientists think it quite possible that life molecules or even organisms have traveled from Mars to Earth or from Earth to Mars as a result of large meteor impacts blowing off material at escape speed from each planet. But who knows how any life organisms might survive the tens or hundreds of thousands of years of travel between the planets?[3] (Such material blown off the planet would not take a direct route from Mars to Earth or from Earth to Mars, but rather wander on complicated orbits until eventually bumping into a planet.)

There is no point debating whether human Mars science goals are better or are achieved more economically with robotic missions. The whole thesis of this book is that we will send humans to Mars because we want them there. Once there, science will be a driver. However, there will be continual debate as to whether looking for life (past/extinct or present/extant) is more of a driver than is preparing for eventual human habitability there. The two goals may be in conflict—generally, human habitation is not good for environmental preservation. In chapter 8, I argued that the question of establishing another home for humans was the primary (if not the only) justification for human Mars missions. I also concluded that humans will never reach beyond Mars because the time scale of human space travel is so slow compared to the time scale of robotic and virtual technology development. This question of time scale is crucial. It is the space race I cited in that chapter: the one where real exploration by humans proceeds slowly while virtual exploration gains momentum.

In this appendix, I considered the elements of human Mars mission design to back up the claim that it is a bigger task (and will take longer) than were the Apollo missions to the Moon, even though they started in the space age infancy. These elements are

* the launch vehicle—bigger than anything we have and we'll need many;
* the entry vehicle—an unsolved design complicated by the very thin Mars atmosphere and the need for precise navigation to bring many elements together on the Martian surface;

* in-space propulsion: a new nuclear reactor in space;
* in situ propellant production: making fuel from Martian indigenous resources reliably for the ascent rocket for return to Earth;
* life support for 1.5 years on the Martian surface. In addition, we will need supplies and life support for the 6–8 month interplanetary trips to and from Mars;
* planetary protection and environmental protection of Mars.

This will take a few decades, but as explained in chapter 9, we can start on the humans-to-Mars path right now with the very exciting venture of moving an asteroid, enabling the first human step within a decade. My optimistic scenario (again, discussed in chapter 9) has that first human mission to Mars occurring in the 2040s. But we'll want several such Mars missions to various locales on the planet; then, and only then, can we start building up the infrastructure for human habitation there. That infrastructure will need to be on the Martian surface and also on the transportation system to serve us going back and forth to Mars. This too will take many decades— bringing us well into the 2200s. By then, virtual exploration with fantastic robotics and information processing will be exploring the universe. But humans will have made it to another world and be a multi-planet species.

Notes

Introduction

1. Thomas O. Paine was the administrator of NASA when the Apollo astronauts landed on the Moon. Following the space shuttle *Challenger* accident (1986) when people were questioning the value and purpose of human spaceflight, he was appointed by President Reagan to lead the National Commission on Space to figure out those issues.

2. My fellow co-founders of The Planetary Society.

3. Douglas Adams, *The Hitchhiker's Guide to the Galaxy* (New York: Harmony Books, 1980).

4. Louis Friedman, *Starsailing: Solar Sails and Interstellar Travel* (New York: Wiley, 1988).

5. The unobservable universe consists of dark matter, dark energy, and who knows what else—perhaps connections to other universes. This is beyond our reach and certainly beyond the scope of this book. Alert readers might be asking how we can observe 45 billion light-years if the age of the universe is only 13.8 billion years. Light can't travel faster than the speed of light, which is by definition 1 light-year per year. But, the universe is expanding; hence, while the light travels from a source, that source is accelerating and flying away from us.

Chapter 1

1. In fact, today's smartphone has about 30,000 times the memory and operates about 1,000 times faster than the computer that guided *Apollo 11* to the Moon.

2. Moore's law, named after Gordon Moore, the founder of Intel Corporation, is more precisely an observation of the size of electronics.

3. Erwin Schrödinger, *What Is Life? The Physical Aspect of the Living Cell* (Cambridge: Cambridge University Press, 1992).

4. "Ah, but a man's reach should exceed his grasp, Or what's a heaven for?" (Robert Browning).

5. Most of the time in this book, "we" refers to humans.

Chapter 2

1. AU is astronomical unit, the mean distance of the Earth from the Sun. It is approximately 93 million miles or 150 million kilometers.

2. Interstellar space, in this definition, does not mean beyond the solar system. The solar system—where the Sun's gravity dominates that from any other star— extends much, much farther, at least to about 100,000 AU.

3. This book is intended for publication in the United States where we sadly seem unable to learn the metric system, used by almost all the rest of the world. Nonetheless, we should—especially when dealing with planetary ideas. Ten kilometers is about 6.2 miles. One trillion is 10^{12} (one followed by 12 zeros), or one million million. One trillion used to be an unimaginably large number, but it is now the unit used for the annual budget in dollars of the United States, and hence we hear it used every day. However, that does not make flying 40 trillion kilometers (the distance to the nearest star) any easier.

4. George Dyson, *Project Orion: The True Story of the Orion Spaceship* (New York: Henry Holt and Co., 2002).

5. Marc G. Millis and Eric W. Davis, eds., *Frontiers of Propulsion Science*, Progress in Astronautics and Aeronautics, vol. 227 (Reston, VA: American Institute of Aeronautics and Astronautics, 2009).

6. Robert L. Forward, *The Flight of the Dragonfly* (New York: Timescape Books, 1984).

7. "Forward" is a nice double entendre denoting the necessary forward thinking for interstellar flight and honoring the aforementioned Dr. Robert Forward.

8. "Project Forward—Beamed Propulsion," Icarus Interstellar, accessed February 20, 2015, http://www.icarusinterstellar.org/projects/project-forward/.

9. *Principium: The Newsletter of the Institute for Interstellar Studies*, Institute for Interstellar Studies, issue 6, Starship Congress Special Issue [2013], http://i4is.org/app/webroot/uploads/files/Principium_Special_Summer_2013.pdf.

Chapter 3

1. To use the title of the Leslie Bricusse and Anthony Newley musical, *Stop the World —I Want to Get Off*, that is what this is. Or more accurately, it is stopping ourselves from orbiting the Sun with the Earth and then restarting ourselves to orbit the Sun in the opposite way.

2. Astronomers use the word "apparition" to refer to a comet becoming visible to Earth during its flight in the inner solar system. The usage of the word obviously goes back to the time before they could be observed by telescopes.

3. See chapter 2 for a discussion of the Voyager missions.

4. Not including Pluto, but Pluto is not a planet—even though it was then.

5. "We" means people of Earth. The figure of 24 men to the Moon includes those who orbited or flew around the Moon as well as those who landed. Three men each went twice.

6. Very recently, my colleague John Logsdon informed me that it was NASA's decision not to continue Apollo. He documents this in his book *After Apollo? Richard Nixon and the American Space Program* (New York: Palgrave Macmillan, 2015).

7. That was in 1980, and it was the reason that Carl Sagan, Bruce Murray, and I formed The Planetary Society to prove there was public interest and political support for space exploration. We, working with many others, forestalled the elimination

of planetary programs from NASA, but still, the 1980s were a dark decade for solar system exploration.

8. As executive director of The Planetary Society, I participated in a bizarre meeting on an early Sunday morning in the Washington mansion of toy heiress Barbara Marx Hubbard with White House personnel and industry leaders who believed they could make this happen.

9. Comet Halley has a period of approximately 76 years and comes through the inner solar system rather fast during that period. Its next visit will be in 2061–2062. Hopefully we'll do a mission then—maybe a rendezvous with a solar sail!

10. Venera-Galley (VEGA) in Russian.

11. The VEGA mission was the first mission of the Soviet Union that was conducted openly, with access to western participants. Under the leadership of its scientific chief, academician Roald Sagdeev, it anticipated the reforms of glasnost and perestroika that preceded the end of the Soviet Union.

12. Andrea Wulf, *Chasing Venus: The Race to Measure the Heavens* (New York: Alfred A. Knopf, 2012).

13. This was similar to the Lunar XPrize initiated nearly 20 years later on the heels of the suborbital XPrize flight success. The Lunar XPrize, unlike the solar sail competition, did garner financial support—principally from Google.

14. Louis Friedman, *Starsailing: Solar Sails and Interstellar Travel* (New York: Wiley, 1988); Jerome L. Wright, *Space Sailing* (Philadelphia: Gordon and Breach Science Publishers, 1992); Colin R. McInnis, *Solar Sailing: Technology, Dynamics and Mission Applications* (New York: Springer, 2004); Giovanni Vulpetti, Les Johnson, and Gregory L. Matloff, *Solar Sails: A Novel Approach to Interplanetary Travel* (New York: Copernicus Books, 2008).

15. The designation of "first solar sail spacecraft" is arguable. *Cosmos 1* in 2005 was the first solar sail flight spacecraft built and launched. Unfortunately, it never flew—the launch vehicle failed before even leaving Earth. The Japanese *IKAROS* certainly deserves credit for the first solar sail flight in 2010. The Russian *Znamya* was the first sail deployed in space, but as noted, it was in the Earth's atmosphere and not a solar sail flight vehicle. NASA's *Nanosail-D* attempted a flight in 2008 also lost with a launch failure and then successfully flew a second spacecraft in 2011. However, like *Znamya*, it was in the Earth's atmosphere and not a solar sail flight vehicle.

16. A great name; *Icarus* (the usual English spelling, but it is a transliteration of the Greek) is of course the legendary Greek who flew too close to the Sun.

Chapter 4

1. Our main focus, however, was advocacy, and by 1984, we were promoting international cooperation as part of that advocacy. In some ways our advocacy of Soviet-American cooperation was as risky as our advocacy of novel technical projects. Both were designed to create a movement for space exploration, not just individual achievements.

2. "Purple pigeons" were so named in the attempt to create exciting and interesting space missions that were distinguished from "grey mice," far more pedestrian

proposals. The Comet Halley solar sail was one of the purple pigeons, as were a Titan probe and Mars rover/sample return.

3. They actually offered two free launches by including a suborbital flight for us to test the sail deployment system. When that flight failed (not our sail test, but the launch vehicle itself), it cost us time and money to deal with the problems, reminding me of the punny aphorism, "there is no such thing as a free launch."

4. A couple of years later, NASA proposed an equivalent mission with about the same objectives and proposed capabilities for about $85 million!

5. Blastoff!'s company goal was to fly a private lunar lander and commercially sell data products from the mission. It attracted a top technical team but ultimately had to be abandoned when the cost of meeting the mission objectives exceeded the amount that could be raised by private investors.

6. Diamandis later went on to fame as the XPrize founder and creator of the Ansari XPrize for the first privately funded human spaceflight—a suborbital flight in 2004.

7. The first *Kosmos* (the usual spelling for the transliteration) flew in 1962. More than 2,000 *Kosmos* spacecraft (both military and civil Earth orbiters) have flown since.

8. Although space is said to begin at an altitude of 100 kilometers above the Earth's surface, true disappearance of the atmosphere does not begin until much higher altitudes. The atmosphere is heated by solar radiation, and the actual altitude where the number of molecules becomes negligibly small varies depending on solar activity—it can be as low as 650 kilometers and as high as 900 kilometers. To properly plan a mission above the atmosphere, one has to account for the solar activity and predict it.

9. While launch failures are part of the space business and occur all too frequently in the United States as well as in Russia (and elsewhere), our bad experience with the Russian company who manufactured that rocket led us to give up on the idea of accepting even free launches on their converted ICBMs. They offered; we refused.

10. No one has given me an explanation of what "D" meant in the *NanoSail-D* name. It was not the fourth spacecraft in a series. I suspect it meant "drag" since the sail was to be deployed in the atmosphere and serve as a brake (rather than as a real sail), using atmospheric drag to de-orbit the spacecraft.

11. In addition to technical innovation, this project shows an even more remarkable innovation by The Planetary Society: international cooperation, working with Americans, Europeans, and Russians.

12. "Smallsat" is a generic name for satellites less than 100 kilograms. "Microsats" are 10–100 kilograms, "nanosats" are 1–10 kilograms, and "picosats" are less than 1 kilogram. These terms are neither grammatically nor mathematically correct in their usage, but they are now conventionally accepted terminology.

13. The donor wishes to remain anonymous.

14. Actually two spacecraft were built—a flight unit and a spare—the latter being cheaper than buying an insurance policy against launch vehicle failure as well as providing additional spacecraft options.

15. I had already stepped down as The Planetary Society executive director in September 2010, passing the mantle to Bill Nye, the Science Guy, a television celebrity famous for science education.

16. Additionally, the *Nanosail-D* second try resulted in an apparently successful flight in 2011 (albeit still within the Earth's atmosphere).

17. "A Mission to Clear Dangerous Debris from Space," University of Surrey, accessed February 20, 2015, http://www.surrey.ac.uk/mediacentre/press/2010/26099_a_mission_to_clear_dangerous_debris_from_space.htm.

Chapter 5

1. DARPA trademarked this title and licensed it to the winner of a study competition that they funded.

2. Noted physicist and long-time interstellar-flight thinker Freeman Dyson has suggested something similar, with biological payloads traveling to and communicating from the stars.

3. J. Craig Venter, *Life at the Speed of Light: From the Double Helix to the Dawn of Digital Life* (New York: Viking, 2013).

4. In the previous chapter we mentioned that the Jet Propulsion Laboratory and NASA Marshall Space Flight Center have already initiated two missions: one to the Moon and one to a near-Earth asteroid.

5. These values are approximate, as the actual temperatures depend on the spacecraft and sail geometry and the time spent in these regions.

6. Pluto has an average orbital distance of 40 AU, but Pluto is not a planet, and because its orbit is very elliptical, it isn't always farther than Neptune.

7. We haven't observed this many; these are predictions based on analyses of celestial mechanics and extrapolations of what we do observe.

8. Dwarf planets could be confused with "minor planets"—the former name for asteroids.

9. 1 AU/year = 4.73 km/sec (about 17,000 km/hr or about 10,500 mi/hr).

10. It is called a "cloud" and there are probably very many objects there. But it is still mostly empty space—the distances between objects will be huge.

11. Some models predict the Oort Cloud to extend farther—to perhaps 100,000 AU. Some models also predict that it might start as near as 2,000 AU.

12. We started to include six of these accelerometers (built by Lumedyne Technologies) on our *LightSail*® spacecraft (chapter 4) but only as an experiment to test how they would work. They were subsequently dropped from the payload after a failed ground test.

13. I tell one of these stories, involving Heinsheimer, in connection with the goal of Mars exploration in chapter 8.

14. We use area to mass ratio (A/m) in units of meters squared per kilogram to describe the spacecraft performance. These are both in units we conventionally visualize. The "characteristic" acceleration of the spacecraft—defined as the acceleration at 1 AU. It equals 0.93 A/m in units of micro-g, that is, millionths of Earth's gravity.

15. E.g., a 100×100 m sail with a 10 kg spacecraft or a 32×32 m sail with a 1 kg spacecraft (a "picosat").

16. Such a craft would be traveling at ~2,500 AU/year, still only 4 percent the speed of light. That speed would cover the distance from Earth to Pluto in about six days! At the interstellar target we are either going to have to slow down a lot using

external light energy or make some very fast measurements. Getting meaningful data once we arrive at our destination is a problem we leave for our grandchildren, or theirs.

17. Johanna Bible, Isabella Johansson, Gary B. Hughes, and Philip M. Lubin, "Relativistic Propulsion Using Directed Energy," Proc. of SPIE, 8876, 887605.

Chapter 6

1. JPL performed a mission study for a Thousand Astronomical Units mission (TAU) several years ago, but with a very large spacecraft powered by nuclear propulsion.

2. This paradox was posed by the famous physicist Enrico Fermi, who reasoned that if there really are so many planets on which life has evolved that, given the time they have had (including the many that evolved faster than did Earth), surely a few of the likely intelligent species would have developed space travel and come to our solar system to investigate. If so, where are they?

3. In addition to brilliant intellects, SETI also attracts brilliant achievers. It was my pleasure to meet both Steven Spielberg, who donated significantly to The Planetary Society's SETI project, and Paul Allen, who donated the start-up funds for the Allen Telescope Array (ATA) of the SETI Institute.

4. Claudio Maccone, *Deep Space Flight and Communications: Exploiting the Sun as Gravitational Lens* (New York: Springer, 2008).

5. The same justification is often used for playing the lottery—if you don't play, you can't win. My guesstimate is that the odds of me winning the lottery are about the same as a radio astronomer finding a beamed intelligent signal. So, it's a question of money. At low cost, SETI is very worthwhile. At high cost, it's a tougher call (and depends on who is paying and who is working).

6. I refer to the Voyager project scientist and former director of JPL, now the CEO of the Thirty Meter Telescope development in Mauna Kea, Hawaii.

7. The radioisotope is usually plutonium.

8. One of the earliest RTGs in the U.S. space program was SNAP-3B in 1961(!). It weighed 2.1 kg with an electrical power output of 2.7 watts. It was only slightly larger than a single CubeSat.

9. Louis Friedman and Darren Garber, *Science and Technology Steps into the Interstellar Medium*, IAC-14,D4,4,3,x22407, Sixty-Fifty International Astronautical Congress, Toronto, Canada, 2014.

Chapter 7

1. I inconsistently, but conventionally, specified light in units of wavelength (nanometers) and sound in units of frequency (hertz, which are cycles per second). All electromagnetic radiation—light, sound, x-rays, radio waves, etc.—is measured by its wavelength or frequency; wavelength equals the speed of light divided by frequency.

2. This was a privately funded venture—in fact, even now the only privately funded instrument to be sent to another planet.

3. Also, to be fair to the robots, many (most) claims of a human "sixth sense" are overblown.

4. A first-generation 3-D printer has now been sent to the International Space Station (ISS). It presages a future of making things in space with diverse uses, from pizza to electronics and mechanical components.

5. "Immersive Space Exploration," JPL, accessed February 20, 2015, http://www .youtube.com/watch?feature=player_embedded&v=UyqjmrPrJ8s.

6. Venter, *Life at the Speed of Light.*

7. One might think of quantum communications (mentioned in chapters 2 and 6) offering speeds for information transfer faster than the speed of light, but applying these purely theoretical physics ideas (like this wormhole travel and others) is, for me, still science fiction.

8. Ray Kurzweil, *The Singularity Is Near: When Humans Transcend Biology* (New York: Viking, 2005).

9. I wrote an essay about the idea presented here, viz. that human spaceflight will not go beyond Mars, and I was accused in various commentaries of being negative and giving up on the future of spaceflight. One headline read, "A Defeatist View of Human Exploration." Time will tell whether human exploration beyond the solar system will take place with or without humans actually in the spaceships. Time will tell, but it won't tell us—it will take centuries before the path is determined.

Chapter 8

1. At the time of this book's writing, the earliest humans to Mars flight that is being discussed by NASA or any other space agency is in the late 2030s and more likely the 2040s. In 2013, an exciting science fiction movie, *Europa Report*, depicted a human-crewed mission to explore Europa and discover life there. That is unlikely to bring the reality of such a human flight any closer, although it might help promote the robotic mission.

2. Some enthusiasts argue that space colonies can be created on giant space stations—for thousands of people maybe, but not for hundreds of millions, let alone billions.

3. Viktor left Russia and eventually ended up working at JPL on Venus balloon missions.

4. The precise numbers are subject to the definitions of what is success and/or failure, what exactly having reached Mars means, and distinctions between different spacecraft on the same mission.

5. In the U.S. less than 0.1 % of the federal budget is spent on Mars exploration.

6. Air conditioning in Arizona and subterranean shopping malls in Minnesota are two examples.

7. Kim Stanley Robinson, *Red Mars* (New York: Bantam Books, 1993); Robinson, *Green Mars* (New York: Bantam Books, 1994); Robinson, *Blue Mars* (New York: Bantam Books, 1996).

8. In addition, private missions might be attempted; several enthusiasts, including SpaceX founder Elon Musk have announced their intent for private human missions to Mars. I would love to see it, but 12 years after the first private spaceflight (Dennis Tito's tourist trip to the space station) and 9 years after the first private space

vehicle (SpaceShip One), we have seen that private developments, especially for really ambitious ventures, will take longer than enthusiasts hope for.

9. As of the end of 2014.

10. Phobos is one of the two Martian moons (Deimos is the other), and hence a mission there counts as a Mars mission—especially so when it is a sample return which will likely bring back Mars dust deposited on its surface.

11. As was done on the Cassini mission with the European probe on the American orbiter.

12. Nothing happened until the government got back in, with NASA rescuing the Lunar Prospector, starting the U.S. Department of Defense Clementine lunar mission, and joining other nations in the International Halley Watch. That also happened with the private effort in England to do a Mars lander, *Beagle 2*. It too had to be rescued by the government (the ESA), although ultimately it failed.

13. Mars missions have about a 50–60 percent success rate, but it's getting better.

Chapter 9

1. After we get to the other side of our metaphorical river, we will need the third development I mentioned: the entry vehicle for carrying a crew into the thin Martian atmosphere.

2. Even just going back to the 1960s type of lunar landing goal would require large increases in the national space budget and more than 20 years. This was what was demonstrated in the now-cancelled Constellation Program.

3. SLS stands for Space Launch System, a new U.S. rocket to reach Earth orbit and, with later upgrades, into the solar system. *Orion* is the crew module for the SLS. Other crew modules are being developed commercially by aerospace companies.

4. Strictly speaking only L4 and L5 are stable points—the other three are far less stable. A variety of physical forces perturbing orbits make these points of more mathematical than physical significance.

5. John Brophy, et al., "Asteroid Return Mission Feasibility Study," AIAA–2011–5665, 47th AIAA/ASME/SAE/ASEE Joint Propulsion Conference and Exhibit, San Diego, California, July 31–August 3, 2011.

6. John S. Lewis, *Mining the Sky: Untold Riches from the Asteroids, Comets and Planets* (Reading, MA: Addison-Wesley Pub. Co., 1996).

7. Retrieving a boulder might not seem quite like retrieving a whole asteroid, but planetary scientist Professor John Lewis of The University of Arizona reminded us that all asteroids are just pieces of other asteroids (as a result of collisions among them).

8. Brophy and Strange work at JPL.

9. In the words of President George H. W. Bush.

10. "Moon, Mars and Beyond" was the slogan of President George W. Bush's Vision for Space Exploration. But as this book demonstrates, for humans, there is no beyond.

11. Why anyone would urge the United States into another space race to the Moon is beyond my understanding. After winning the first one, there cannot be any advantage in risking loss on a second one.

12. Despite many efforts to argue otherwise, no nation has ever found a military advantage for humans in space, let alone on the Moon.

13. Apollo 11 astronaut Buzz Aldrin has proposed this also.

14. Parts of this chapter dealing with the Asteroid Redirect Mission (ARM) are taken from a paper written by me and former astronaut Thomas Jones for the 2013 International Astronautical Congress. The work was done in a study sponsored by the Keck Institute for Space Sciences (KISS).

Chapter 10

1. Though, possibly, with still many catastrophic ones.

2. Proxmire, 1915–2005, was a critic of space program funding.

3. Word parsing is possible here. "To Mars" as opposed to "on Mars" was actually stated by President Obama to be a goal for the 2030s—the implication being a circumnavigating flight around Mars, or a Mars orbiter, or perhaps even a Martian moon (Phobos or Deimos) landing in the 2030s. The big expense and extra requirement of entry, descent, and landing are then deferred until after we achieve the interplanetary flight and life-support requirement of the Earth-Mars round trip. I think this makes sense—former NASA administrator Mike Griffin suggested exactly such an approach to The Planetary Society in a study before he became NASA administrator and succumbed to the nationalistic thinking in favor of a lunar base.

4. This happened with the lunar base goal adopted by the George W. Bush administration in the Constellation Program. Originally it was a 15-year goal, but by the time it was cancelled, the goal had slipped a decade and was further away (18 years) than when it started. Also, its goal, a lunar base basically creating a self-described "Apollo on steroids," was not so publicly engaging.

Appendix

1. Here, and throughout this section, "we" refers to humans from Earth.

2. Designing faster missions (with stay times of 20–30 days) is in principle possible, but it requires a great deal more energy (propulsion) for both legs of the mission—too much, based on current planning.

3. While at The Planetary Society, I conceived the Living Interplanetary Flight Experiment (LIFE), to test the hypothesis of interplanetary organism space travel. We packaged 10 microorganisms in a small capsule to place on a spacecraft going on a multi-year round-trip planetary trajectory with the intent of examining the organisms when they returned back to Earth. Our first (and thus far, only) attempt to fly this on the Russian Phobos sample return mission was thwarted when the launch vehicle for that mission failed.

Index

About the Author

Louis Friedman received his PhD from the Aeronautics and Astronautics Department of the Massachusetts Institute of Technology in 1971. He co-founded The Planetary Society in 1980 with Carl Sagan and Bruce Murray and served as its executive director for 30 years. Prior to The Planetary Society, he was leader of Advanced Programs at the NASA Jet Propulsion Laboratory. He was a Congressional Science Fellow on the U.S. Senate Committee on Commerce, Science and Transportation. After retiring from The Planetary Society, he led studies on the Asteroid Redirect Mission and on Exploring the Interstellar Medium at the Keck Institute for Space Studies at the California Institute of Technology. He currently serves on the NASA Innovative Advanced Concepts (NIAC) External Council. He is a member of the American Astronautical Society, the Division for Planetary Sciences of the American Astronomical Society, Sigma Xi, and the Explorers Club and a Fellow of the British Interplanetary Society, the American Institute of Aeronautics and Astronautics, and the American Association for the Advancement of Science. In addition to giving numerous lectures, Dr. Friedman is author of *Starsailing: Solar Sails and Interstellar Travel* (Wiley, 1988).